Nelson Maths

W0036063

4

This book belongs to:

_____

Workbook

Karen Morrison
Lisa Greenstein

OXFORD
UNIVERSITY PRESS

OXFORD
UNIVERSITY PRESS

Great Clarendon Street, Oxford, OX2 6DP, United Kingdom

Oxford University Press is a department of the University of Oxford.

It furthers the University's objective of excellence in research, scholarship, and education by publishing worldwide. Oxford is a registered trade mark of Oxford University Press in the UK and in certain other countries.

First published 2022

British Library Cataloguing in Publication Data

Data available

ISBN: 978-1-382-01030-6

1 3 5 7 9 10 8 6 4 2

Paper used in the production of this book is a natural, recyclable product made from wood grown in sustainable forests. The manufacturing process conforms to the environmental regulations of the country of origin.

Printed in Great Britain by Bell and Bain Ltd, Glasgow

Acknowledgements

The publisher and authors would like to thank the following for permission to use photographs and other copyright material:

Cover: Matthieu Nivesse.

Artwork by Aviel Basil, Q2A Media, Alan Rogers, Pantek Media, and OKS Prepress.

Every effort has been made to contact copyright holders of material reproduced in this book. Any omissions will be rectified in subsequent printings if notice is given to the publisher.

Contents

Think maths

What do you think?

1. Read the thoughts in the table below. These can limit your brain.

2. Write down what you could think instead. There are some ideas in the clouds under the table.

Instead of this ...	Try thinking this ...
I can't do this.	
This is too hard.	
I'm not good at maths.	
I'll never be smart.	
My idea didn't work.	
I made a mistake and now I can't do it.	

Mistakes are great – they help me to learn better.

I'm going to figure this out and get better at it.

What am I missing or not seeing here?

This might take time and effort.

I'm going to train my brain to do maths.

3. Can you think of any other thoughts which can limit your brain? Write two in the table.

4. Write what you could think instead of these.

➡ *Pupil Book page 5*

Maths helps your brain grow

1 Complete the statements about your brain.

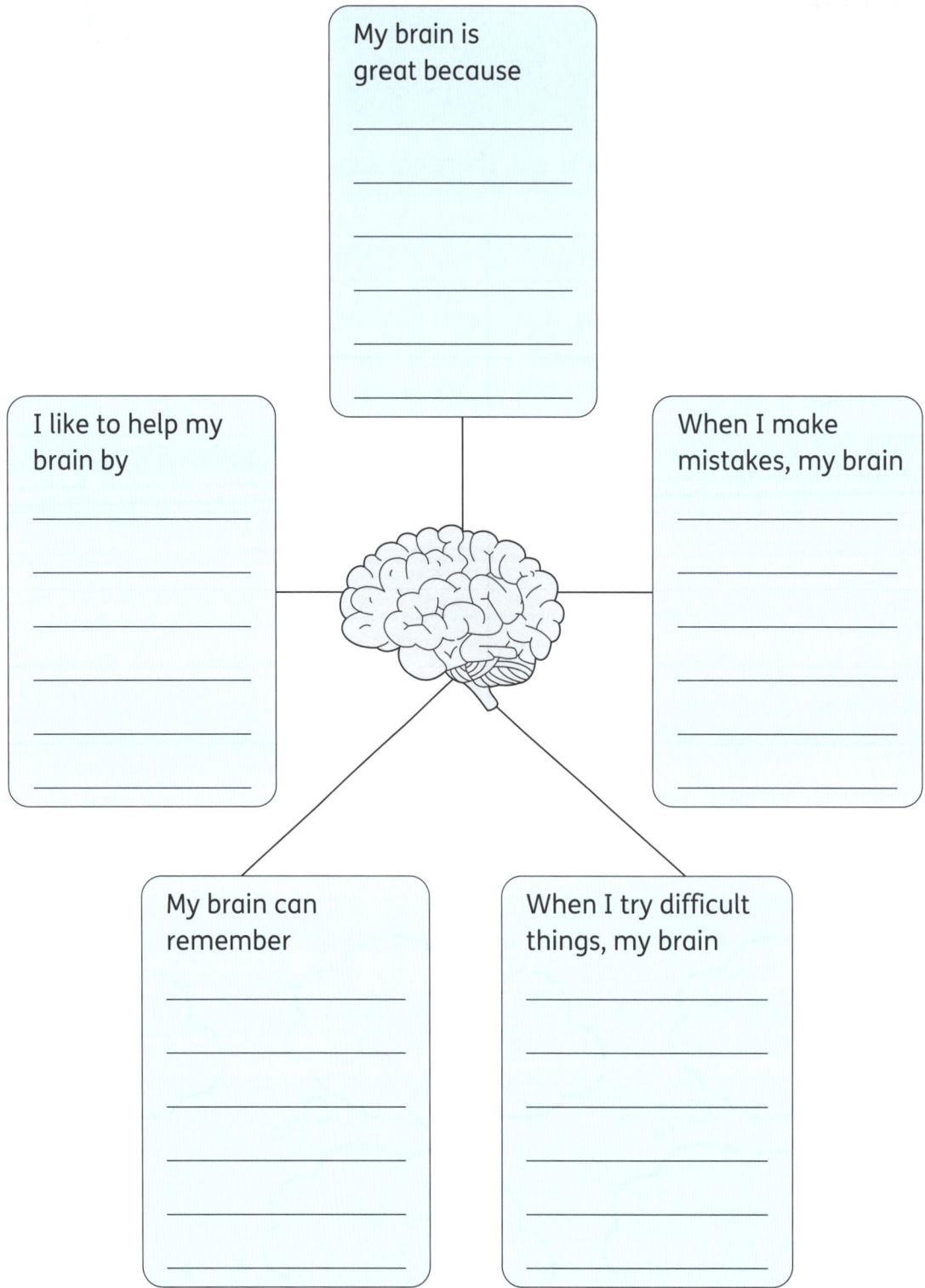

My brain is
great because

_____

_____

_____

_____

_____

_____

I like to help my
brain by

_____

_____

_____

_____

_____

When I make
mistakes, my brain

_____

_____

_____

_____

_____

My brain can
remember

_____

_____

_____

_____

_____

When I try difficult
things, my brain

_____

_____

_____

_____

_____

2 Share your ideas with a partner.

➡ *Pupil Book page 6*

Numbers and place value

Revisit place value

1

6   1   7   2

Use three of the digits above to make:

a the smallest possible 3-digit number _____

b the greatest possible 3-digit number _____

c a 3-digit number between the two numbers you made _____

2

3   8   9   6

Use three of the digits above to make:

a the smallest possible 3-digit number _____

b the greatest possible 3-digit number _____

c a 3-digit number between the two numbers you made _____

Problem solving

3 Use the information to work out each number.

a	It has 24 tens and is 11 away from 254.	
b	If it was 9 greater it would be 60 tens.	
c	If 7 tens were added to it, it would be 400.	
d	It needs 10 tens to make it 999.	
e	It is 5 tens more than 470.	

Use diagrams or models if you need to.

➡ *Pupil Book page 8*

Place value to thousands

1 Write the value of the underlined digit in each number. The first one has been done for you.

a 5<u>2</u>40 __200__   b 1<u>0</u>98 _____   c <u>1</u>609 _____

d <u>3</u>182 _____   e 80<u>5</u>6 _____   f <u>7</u>484 _____

g <u>6</u>179 _____   h 2<u>1</u>47 _____   i 97<u>6</u>2 _____

2 Write each of these numbers using digits. The first one has been done for you.

a Two thousand, eight hundred and forty-three __2843__

b Six thousand and sixty-five _____

c Eight thousand and fifteen _____

d Seven thousand, two hundred and twenty _____

e Four thousand and four _____

3 Find five different 4-digit numbers from a newspaper, magazine or website. Copy them into the table and write them in words.

My numbers	In words

➡ *Pupil Book page 9*

Expanded form

1 Write each number in expanded form. The first one has been done for you.

a 5792 = 5000 + 700 + 90 + 2

b 3650 = _____

c 8275 = _____

d 1960 = _____

e 2009 = _____

f 6090 = _____

2 Here are some numbers with their expanded form. Fill in the missing digits and numbers. The first one has been done for you.

a 1 | 5 | 43 = | 1000 | + 500 + 40 + 3

b [ ] 412 = 3000 + 400 + [ ] + 2

c 4 [ ] 88 = 4000 + [ ] + 8

d 87 [ ][ ] = [ ] + 700 + 90 + 9

e [ ][ ] 00 = 9000 + 200

f 79 [ ] 9 = [ ] + 900 + 90 + 9

3 Use digits to write the number that is equivalent to:

a four thousands, three hundreds and eight ones _____

b two thousands, five tens and four ones _____

c eight thousands and four hundreds _____

d three thousands and twelve hundreds _____

e five thousands, four hundreds, fourteen tens and five ones _____

f nine hundreds, nineteen ones and one thousand _____

➡ *Pupil Book page 10*

Compare and order numbers

1 Write the numbers shown by letters **a** to **f** on the first number line. The first one has been done for you.

2 Here is a set of numbers.

3000
8000
6500
2150
7800
9000
9999

Draw arrows on the second number line to show the approximate position of each number.

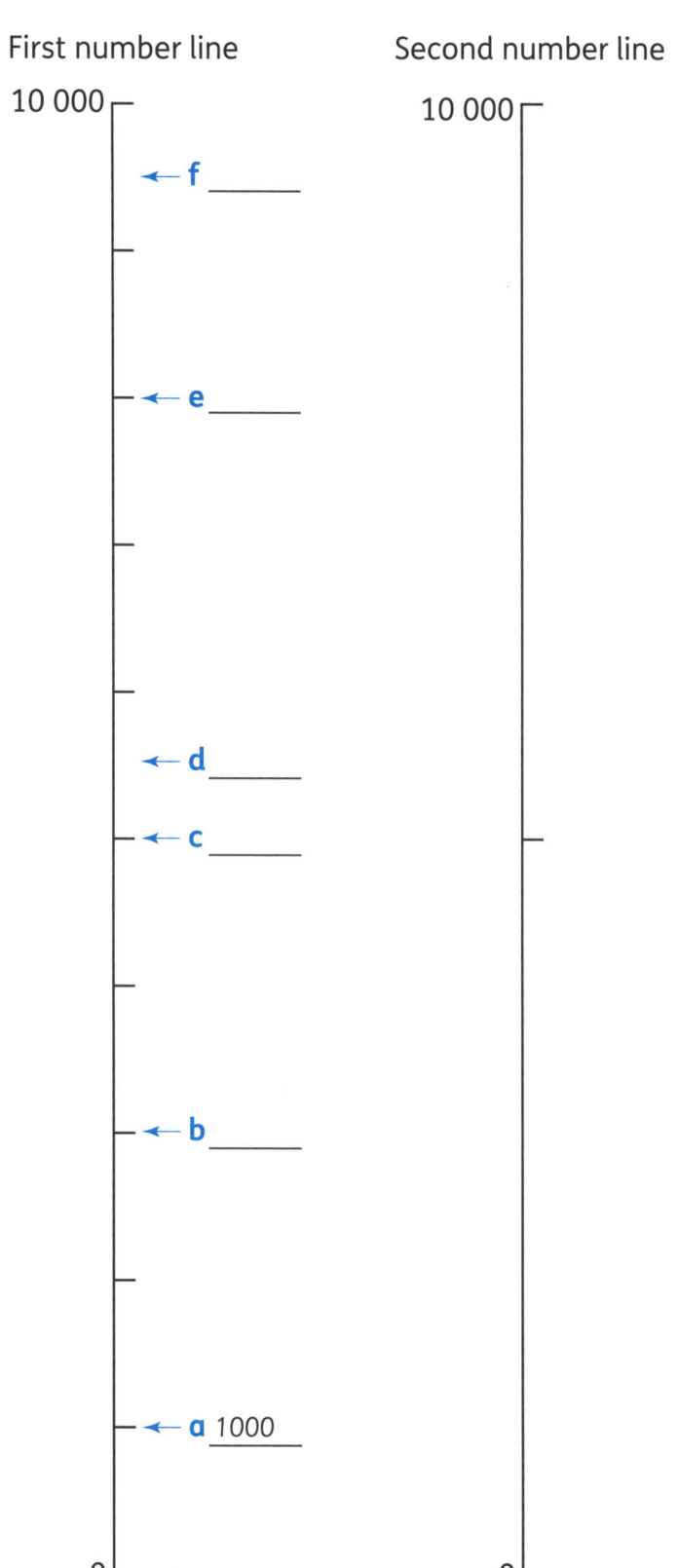

First number line

Second number line

10 000

10 000

← f _____

← e _____

← d _____

← c _____

← b _____

← a 1000

0

0

➡ *Pupil Book page 11*

Use <, > and = to compare and order numbers

1 Fill in <, > or = between each pair of numbers. The first one has been done for you.

a 55 < 129

b 212 ☐ 110

c 10 ☐ 0

d 20 ☐ 200

e 200 ☐ 30

f 195 ☐ 98

g 957 ☐ 599

h 199 ☐ 240

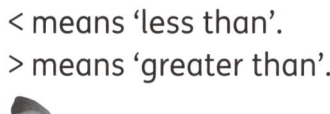

< means 'less than'.
> means 'greater than'.

2 Fill in <, > or = between each pair of quantities. The first one has been done for you.

a double 10 > half of 20

b 100 hours ☐ 2 days

c 600 m ☐ 1 km

d 400 cm ☐ 2 m

e 90 seconds ☐ $1\frac{1}{2}$ minutes

f 500 g ☐ $\frac{1}{2}$ kg

g 7 cups ☐ 2 litres

h 2500 ml ☐ 2.5 l

3 Fill in <, > or = between each pair of calculations. The first one has been done for you.

a 100 − 50 > 200 − 180

b 40 × 2 ☐ 100 ÷ 2

You will need to do the calculations to help you decide.

c 25 + 10 ☐ 35 − 5

d 4 × 4 ☐ 3 × 2

e 9 × 3 ☐ 6 × 4

f 28 ÷ 4 ☐ 30 ÷ 5

g 10 + 10 + 5 + 8 ☐ 25 − 4

h 6 tens ☐ 1 hundred

i half of 18 ☐ 10 − 2

j $\frac{1}{4}$ of 20 ☐ $\frac{1}{2}$ of 10

➡ *Pupil Book page 12*

Compare and order numbers

1 The table shows the names and heights of the highest mountains in different countries. Use information from the table to answer the questions.

Country	Mountain	Height in metres
Antarctica	Vinson Massif	4892
Argentina	Aconcagua	6962
China/Nepal	Mount Everest	8848
Indonesia	Puncak Jaya	4884
New Zealand	Aoraki (Mount Cook)	3754
Russia	Mount Elbrus	5642
Tanzania	Kilimanjaro	5892
United States of America	Mount McKinley	6194

a Which is the highest mountain? _____

b In which country is the third highest mountain? _____

c Which is the only mountain that is less than 4000 metres? _____

d Four of the heights have 8 in the hundreds place. Write these heights in order from lowest to highest.

_____ _____ _____ _____

Problem solving

2 Read the information and work out what the number could be.

- The number is greater than 100 but less than 1000.

- The digit in the tens place is 1 greater than the digit in the hundreds place.

- The number has the maximum possible number of tens.

- Two digits are the same but they aren't next to each other. _____

3 At a school fete, people are asked to guess how many beads there are in a jar. The guess closest to the actual number wins a prize.

The top four guesses are:

1532   1509   1511   1498

There are 1522 beads in the jar. Which guess wins the prize? _____

➡ *Pupil Book page 13*

Round numbers to the nearest 10 and 100

1. Round each number to the nearest 10 and to the nearest 100. The first one has been done for you.

Number	To the nearest 10	To the nearest 100
317	320	300
233		
671		
894		
2185		
3347		

2. Circle the numbers in the box that will round to 1300.

Underline the numbers in the box that will round to 1400.

Round any remaining numbers in the box to the nearest 100. Write the rounded number below each number.

1456	1399	1045	1599	1302	1350
1428	1528	1889	1010	1309	1480
1375	1357	1328	1488	1444	1333
1555	1455	1299	1250	1245	1899

Problem solving

3. A news report says that approximately 1500 people attended an art exhibition.

    a   If the number of people who attended was rounded to the nearest 100, what is the smallest number of people who could have attended?

    _____

    b   If the number of people who attended was rounded to the nearest 10, what is the greatest number of people who could have attended?

    _____

➡ *Pupil Book page 15*

Round numbers to the nearest 1000

Round each number in a box to the nearest thousand.

Join it to the nearest thousand with a line.

The first one has been done for you.

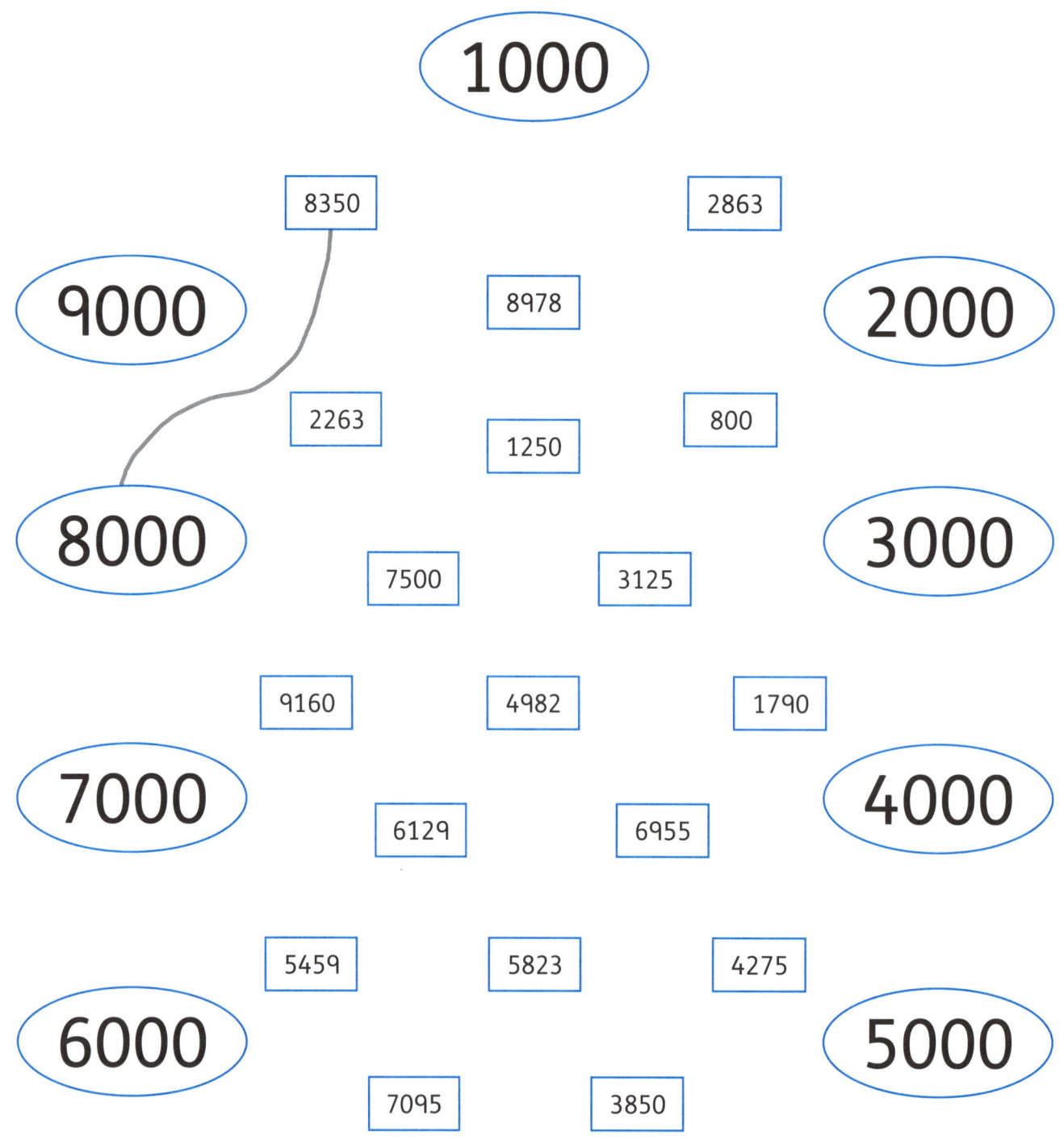

➡ *Pupil Book page 16*

Roman numerals

I	V	X	L	C	D	M
1	5	10	50	100	500	1000

1 Complete the table. Fill in the missing numbers or Roman numerals.

Roman numeral	II		VI			L	LXX	LXXXIX	XCIX	
Number		5		8	10					100

2 Arrange these Roman numerals in order from smallest to greatest value.

XI   XXIV   CX   XC   IX   LV   III

_____ _____ _____ _____ _____ _____ _____

3 The Romans built very good roads. Some of them still exist today. The Romans placed mile markers along the roads so people could work out how far they had travelled.

These mile markers have been placed every 5 miles. Write the correct Roman numerals on each.

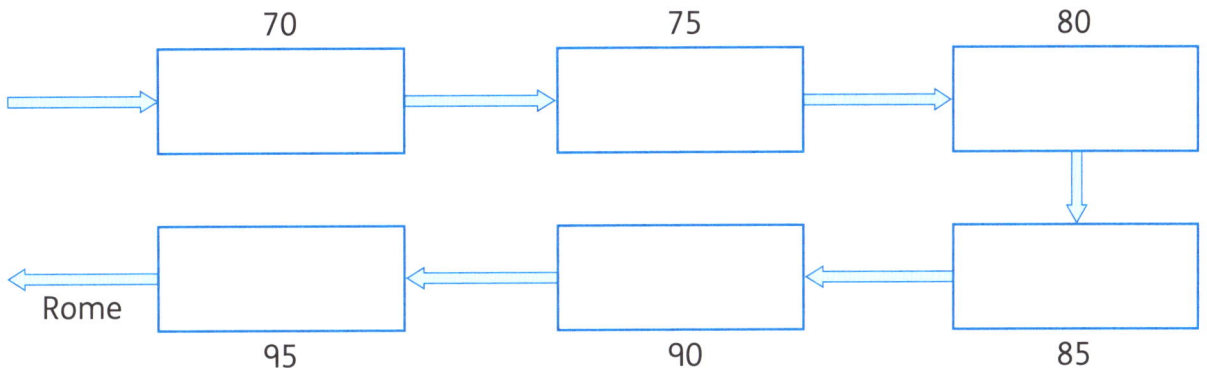

4 Complete these speed limit signs for Roman chariots. Write the speed limits in Roman numerals on these signs.

a 9 miles per hour

b 7 miles per hour

c 2 miles per hour

➡ *Pupil Book page 19*

2D shapes

Shapes and their properties

1. Here are some shape names:
   triangle quadrilateral pentagon hexagon heptagon octagon

   Write the name of each shape. The first one has been done for you.

   a

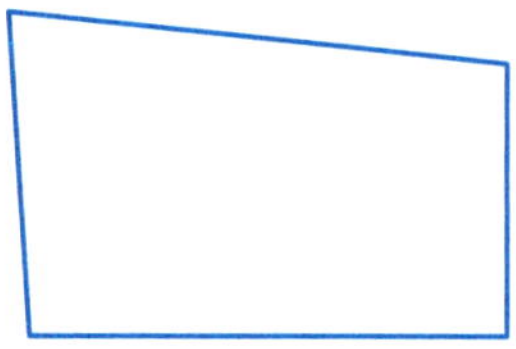

   quadrilateral

   b

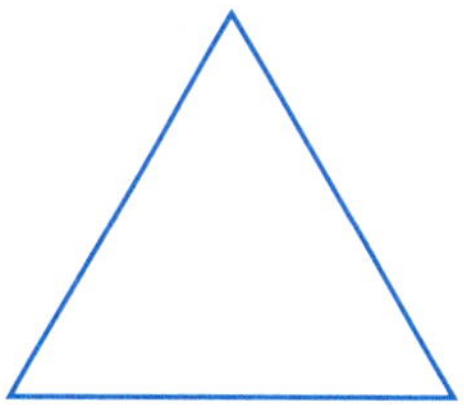

   _____

   c

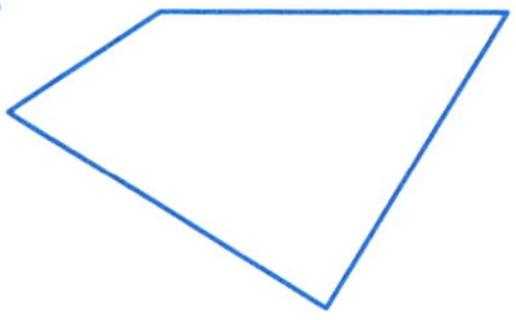

   _____

   d

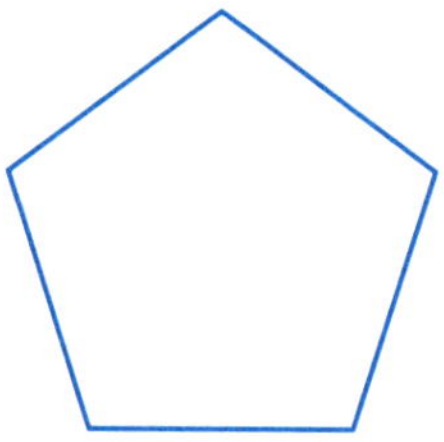

   _____

   e

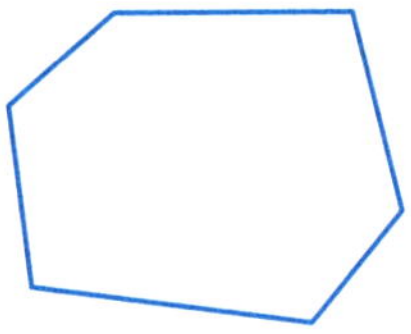

   _____

   f

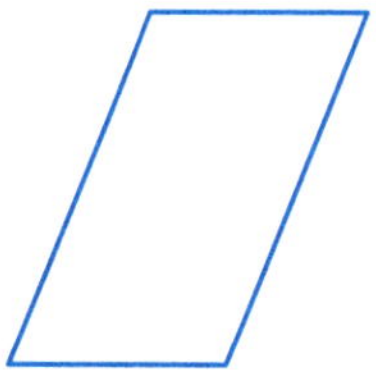

   _____

   g

   _____

   h

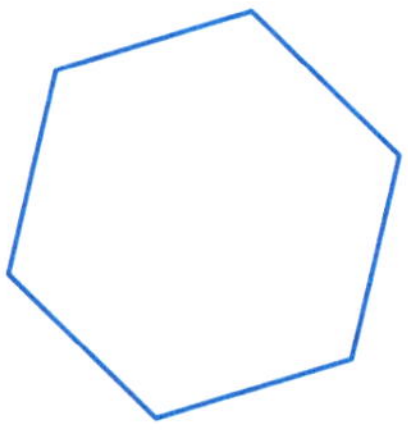

   _____

   i

   _____

   j

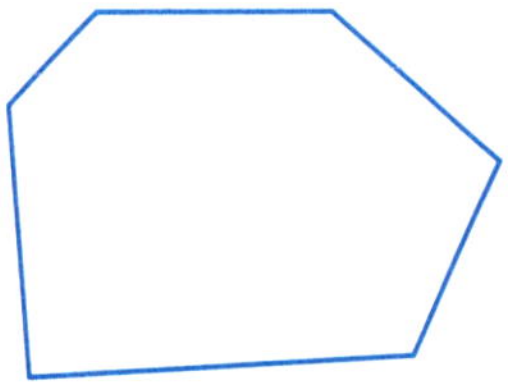

   _____

   k

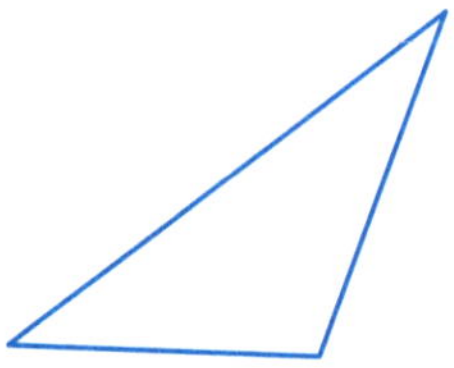

   _____

   l

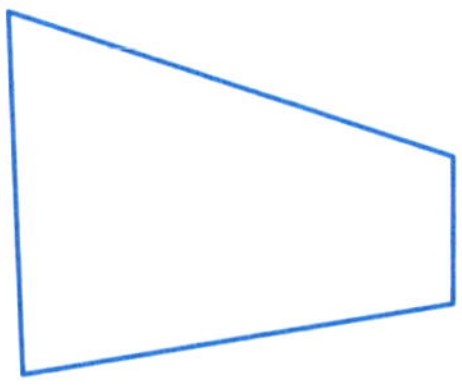

   _____

2. Look at question 1. Decide whether each shape is a regular polygon or an irregular polygon. Write the letters of the shapes in the correct columns in the table.

Regular polygons (all sides the same length, all angles the same size)	Irregular polygons

➡ *Pupil Book page 21*

More shapes and their properties

Complete the table by drawing the shape and writing in the number of sides and angles.
The first one has been done for you.

Use a dictionary or other source to find out about any shapes you don't know.

Name and picture	Number of sides	Number of angles
triangle	3	3
square		
pentagon		
hexagon		
heptagon		
octagon		
nonagon		
decagon		

➡ *Pupil Book page 21*

Quadrilaterals

1 Use 3 rows of 3 dots to draw a quadrilateral. There are 16 different possible shapes. How many can you draw?

These are different.

These are not different. They are just in different positions.

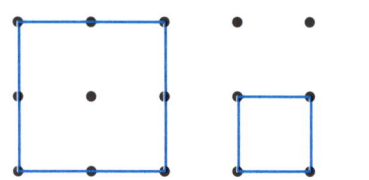 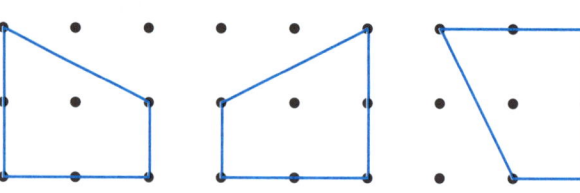

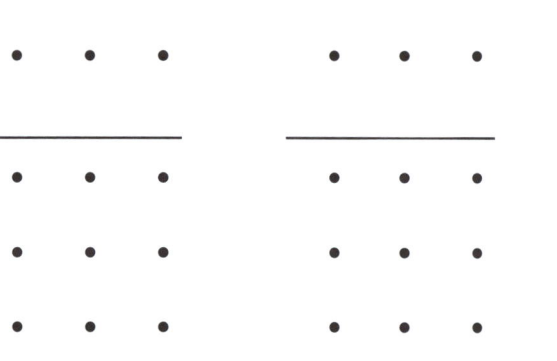

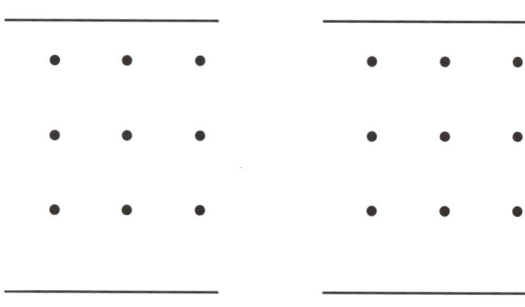

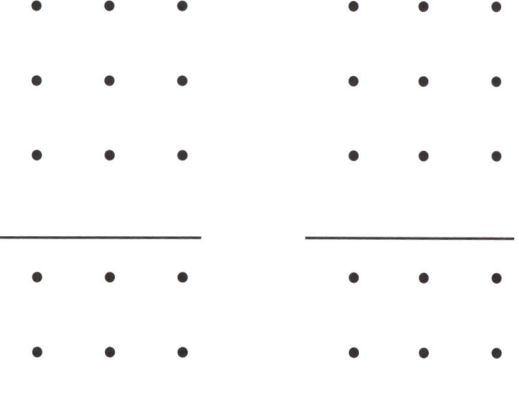

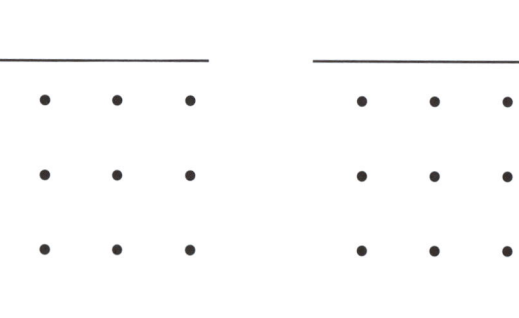

2 Write the names under each of the quadrilaterals you have made, if you know them.

3 Mark all the right angles on each quadrilateral. Mark the right angles like this:

➡ *Pupil Book page 23*

Time

Analogue and digital clocks

Complete each set to show the time in three ways: on an analogue clock, in words and on a digital clock.

1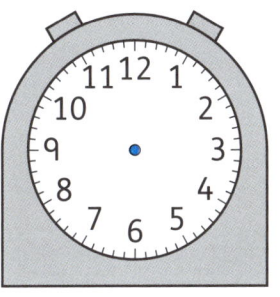

Three o'clock

[:]

2

01:30

_____

3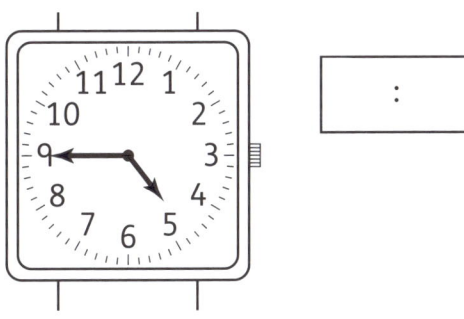

[:]

_____

4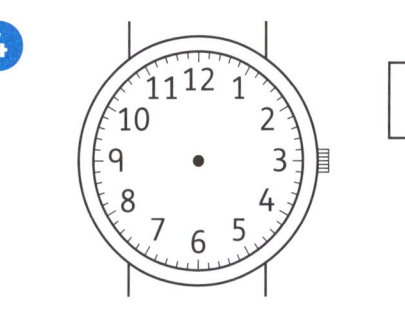

Twenty past eleven

[:]

5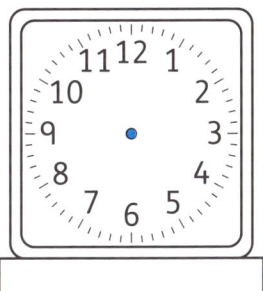

03:15

6

[:]

_____

7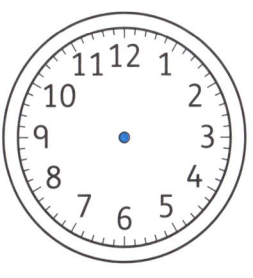

Ten past eight

[:]

8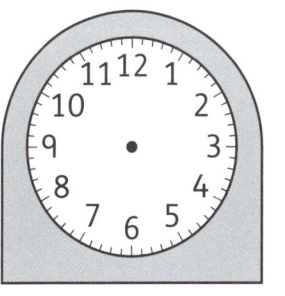

Twelve o'clock

[:]

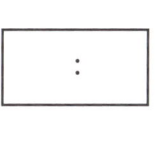 *Pupil Book page 27*

24-hour notation

Show each time on both the analogue clock face and the digital 24-hour clock face.
Remember that 24-hour time always uses four digits.

The first one has been done for you.

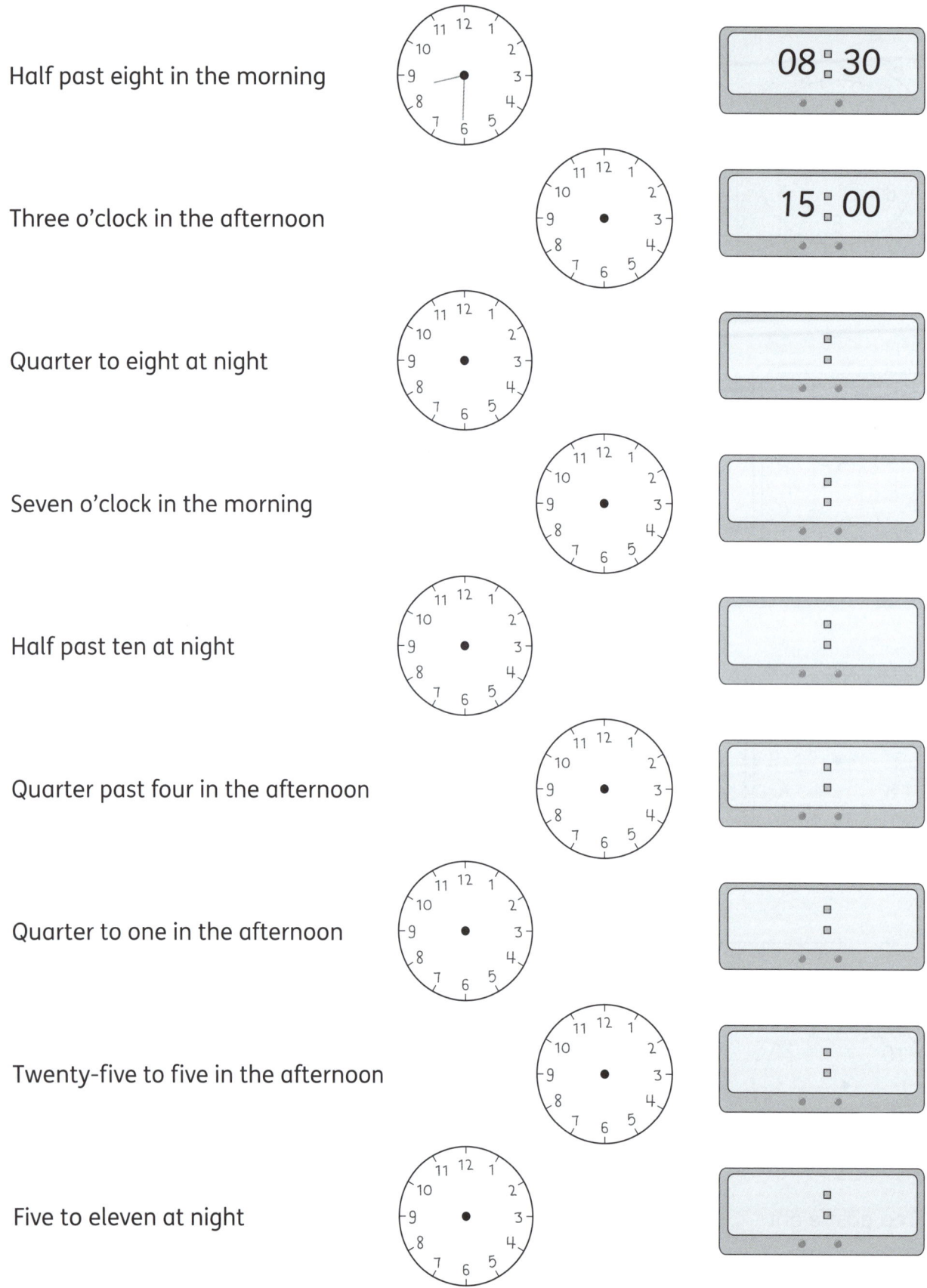

Half past eight in the morning

08 : 30

Three o'clock in the afternoon

15 : 00

Quarter to eight at night

Seven o'clock in the morning

Half past ten at night

Quarter past four in the afternoon

Quarter to one in the afternoon

Twenty-five to five in the afternoon

Five to eleven at night

➡ Pupil Book page 28

Timetables

This timetable shows eight stations on a metro line and the time the train arrives at each one.

Timetable Metro Route 228	
Sungai Buloh	06:10
Surian	07:55
Muzium Negara	09:30
Pasar Seni	10:45
Merdeka	11:00
Bukit Bintang	11:30
Maluri	12:30
Kajang	13:15

1 Where does the train arrive at:

    a half past 12   _____

    b quarter past 1   _____

    c half past 9   _____

    d quarter to 11   _____

2 On public holidays, the train starts later but it still takes the same time between stations. Complete the public holiday timetable below.

Timetable Metro Route 228 – Public Holidays	
Sungai Buloh	07:40
Surian	
Muzium Negara	
Pasar Seni	
Merdeka	
Bukit Bintang	
Maluri	
Kajang	

3 How long will you spend on the metro if you get on at the first stop and get off at the last stop?

_____

➡ *Pupil Book page 29*

More time conversions

1 Fill in a suitable unit of time for each definition in the table. The first one has been done for you.

Unit	Definition
day	a period of 24 hours
	12 months
	$\frac{1}{60}$ of a minute
	365 days
	7 days
	60 seconds

2 Draw lines to match equivalent times. One has been done for you.

2 hours	48 hours	$3\frac{1}{2}$ days	730 days	3 weeks
36 months	300 minutes	6 weeks	3 years	120 seconds
21 days	84 hours	120 minutes	24 months	5 hours
2 days	2 minutes	42 days	49 days	7 weeks

3 Convert these units of time. Show your working.

a 1 day to minutes _____

_____

b 1 hour to seconds _____

_____

c 1 day to seconds _____

_____

d 420 seconds to minutes _____

_____

➡ *Pupil Book page 32*

Decimals

Decimal place value

1 Rewrite each fraction in decimal notation. The first one has been done for you.

$\frac{1}{10} =$ _0.1_ $\qquad$ $\frac{2}{10} =$ _____ $\qquad$ $\frac{3}{10} =$ _____ $\qquad$ $\frac{4}{10} =$ _____ $\qquad$ $\frac{5}{10} =$ _____

$\frac{6}{10} =$ _____ $\qquad$ $\frac{7}{10} =$ _____ $\qquad$ $\frac{8}{10} =$ _____ $\qquad$ $\frac{9}{10} =$ _____ $\qquad$ $\frac{10}{10} =$ _____

2 Complete each number line.

a
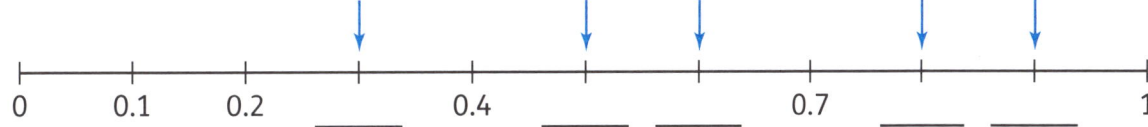

b
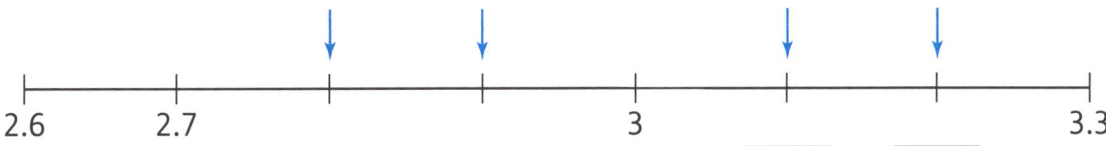

c

d
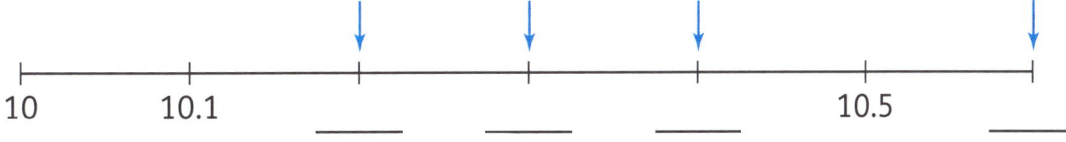

e
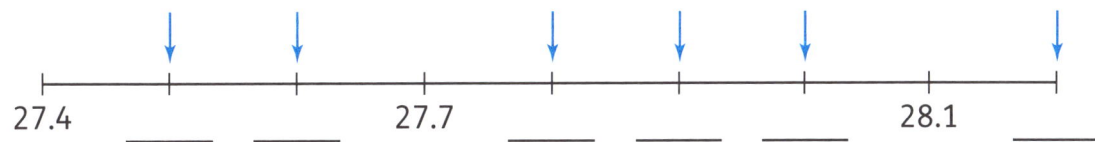

3 Shade the shapes to show the given decimal.

a 0.5

b 0.7
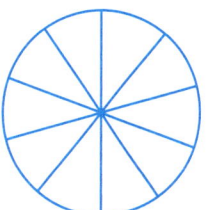

c 0.1

d 1.5
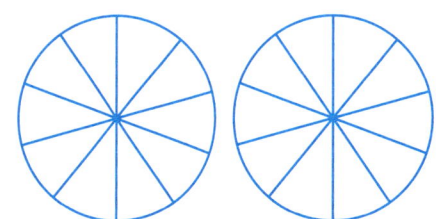

➡ *Pupil Book page 34*

Count in tenths and hundredths

1 This number line is marked in tenths.

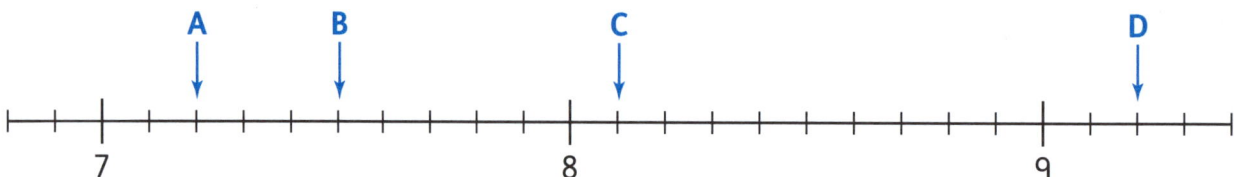

Arrow A is pointing to 7.2.

Which numbers are these arrows pointing to?

a Arrow B _____   b Arrow C _____   c Arrow D _____

2 Show these numbers on the number line below.

a 6   b 4.2   c 3.6   d 4.9   e 3.1

Problem solving

3 These diagrams show sections of different charts marked in hundredths.

Fill in the missing numbers on each diagram.

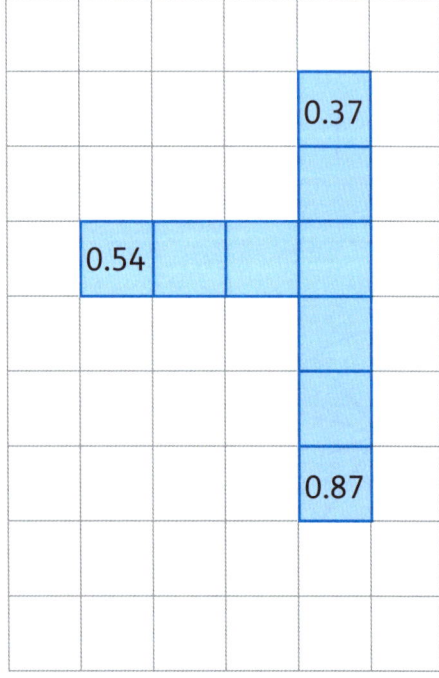

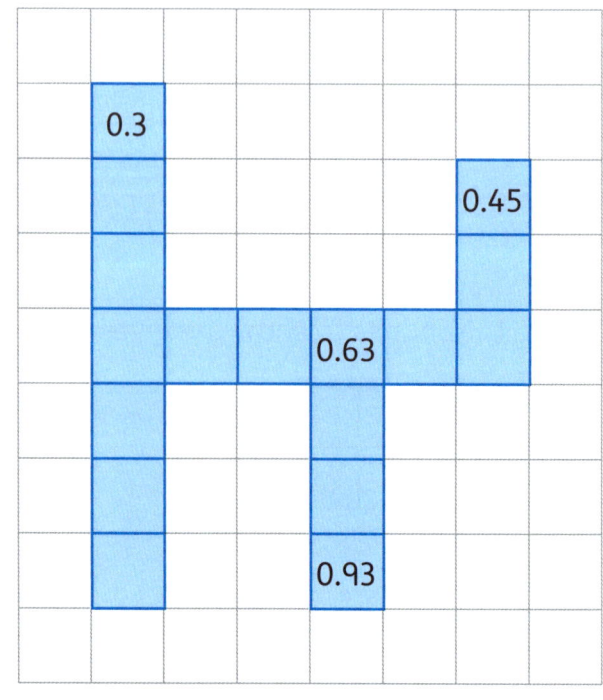

➡ *Pupil Book page 36*

Compare and order decimals (1)

This is a number line with the numbers missing.

Write the correct letter next to each decimal number.

Here is an example.

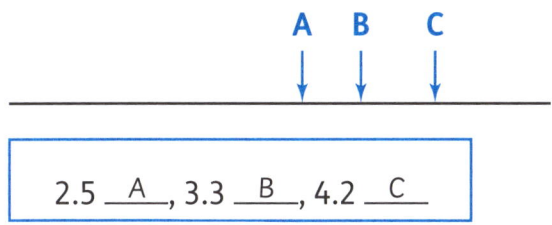

2.5 __A__, 3.3 __B__, 4.2 __C__

A, B and C show the positions of three decimal numbers on a number line.

---

1 Write the correct letter next to each decimal number.

a

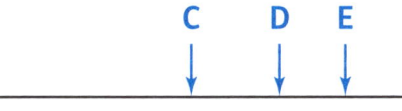

5.3 _____, 7.4 _____, 6.5 _____

b

9.3 _____, 1.2 _____, 8.7 _____

c

20.5 _____, 26.1_____, 15.7 _____

d

25.3_____, 10.5_____, 11.2 _____

e

1.3 _____, 4.9 _____, 2.5 _____

f

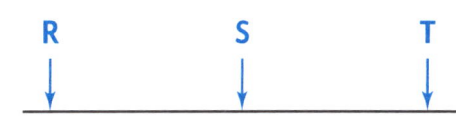

16.8 _____, 8.3 _____, 25.6 _____

g

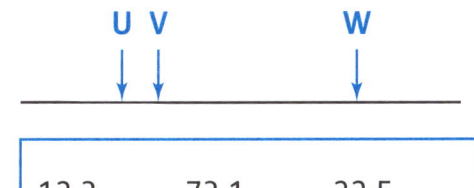

12.3 _____, 73.1 _____, 22.5 _____

h

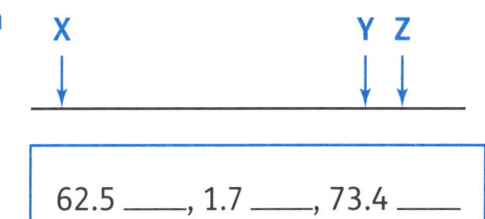

62.5 _____, 1.7 _____, 73.4 _____

➡ *Pupil Book page 38*

Compare and order decimals (2)

1 Circle the smallest amount in each set.

Underline the greatest amount.

a   8.2 ℓ   7.7 ℓ   6.5 ℓ   9.6 ℓ   7.2 ℓ

b   1.2 m   2.1 m   2.5 m   1.5 m

c $4.58 $4.85 $5.48 $5.84

d $12.34 $1.24 $12.00 $123.40

e   14.0 kg   14.4 kg   0.14 kg   0.04 kg

f   12.35 m   123.0 m   123.5 m   1.25 m

g   5.51 kg   3.75 kg   7.35 kg   5.73 kg

h   2.37 kg   2.73 kg   2.07 kg   2.3 kg

2 Rewrite each set of decimals in order from smallest to greatest.

a 0.94   0.95   0.59   _____   _____   _____

b 3.13   2.33   3.3   _____   _____   _____

c 2.05   2.55   5.55   _____   _____   _____

3 Fill in <, = or > to make each statement true.

a 13.8 ☐ 12.8   **b** 0.3 ☐ 3.0

c 12.01 ☐ 12.5   **d** 1.6 ☐ 1.63

e 5.44 ☐ 5.45   **f** 0.9 ☐ 0.90

➡ *Pupil Book page 38*

Measures and money

Measuring instruments

Some instruments that are used to measure length.

There may be more than one correct answer for some items.

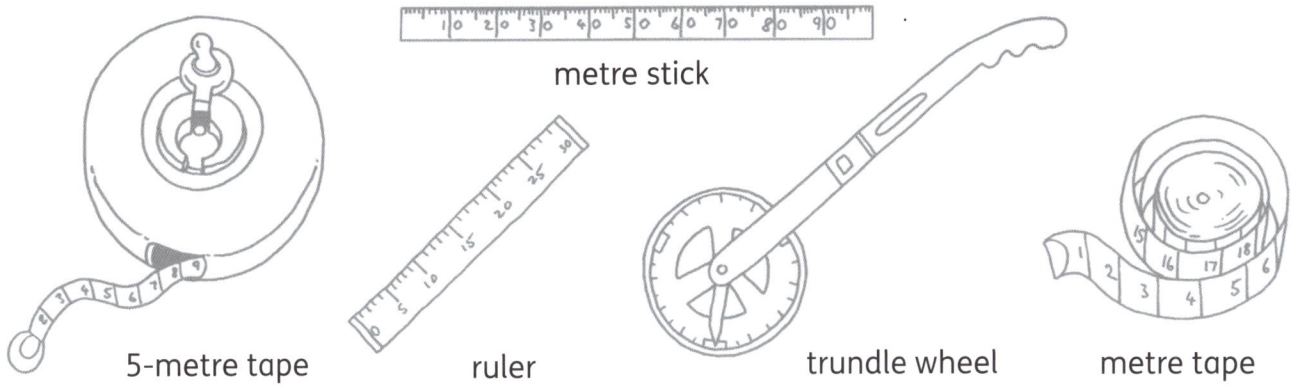

metre stick

5-metre tape   ruler   trundle wheel   metre tape

Which instrument would you use to measure the items in the table?

Tick (✓) the correct columns in the table for each item. Write m, cm or km in the Units column.

Item	Measuring instrument					Units
	Ruler	Trundle wheel	5-metre tape	Metre stick	Metre tape	
basketball court						
pencil						
path						
a belt						
your friend's height						
distance around a tree trunk						
width of a computer screen						

➡ *Pupil Book page 41*

Measure in centimetres and millimetres

Using millimetres on a ruler helps you measure more accurately.

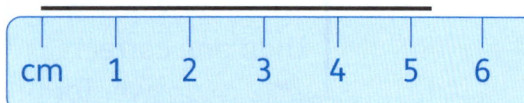

This line is 5 cm to the nearest centimetre.

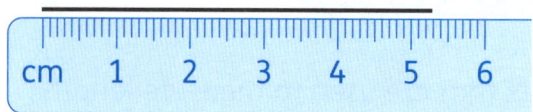

Using a more accurate ruler, the line is about 5 cm 3 mm long.
It is 5 cm 3 mm to the nearest millimetre.
5 cm 3 mm can be written as 5.3 cm.

---

1 Measure these blue lines to the nearest millimetre.

Write their measurements in two ways.

a

_____ cm _____ mm

_____ cm

b

_____ cm _____ mm

_____ cm

c

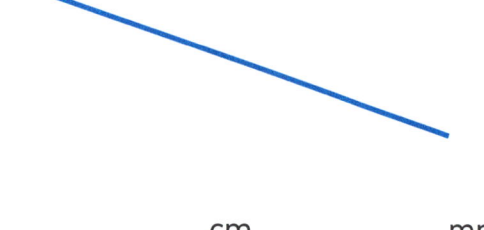

_____ cm _____ mm

_____ cm

d

_____ cm _____ mm

_____ cm

e

_____ cm _____ mm

_____ cm

f Draw a line that is 3 cm and 9 mm long.

➡ *Pupil Book page 42*

Estimate and calculate amounts of liquid

1 Shade each container to show where the level of the liquid will be if you add 150 ml to each one.

Write the total amount of liquid in each jug in millilitres and in litres.

1000 ml = 1 ℓ
500 ml = 0.5 ℓ
250 ml = 0.25 ℓ

a

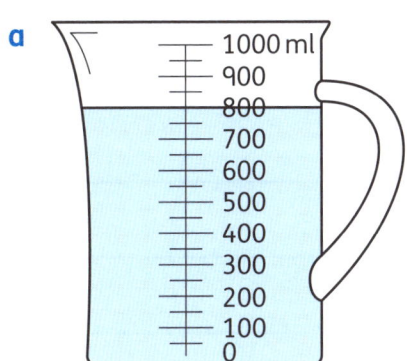

_____ ml

_____ ℓ

b

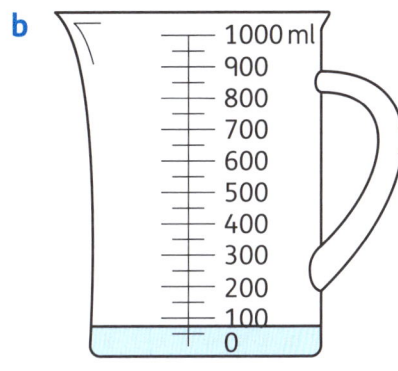

_____ ml

_____ ℓ

c

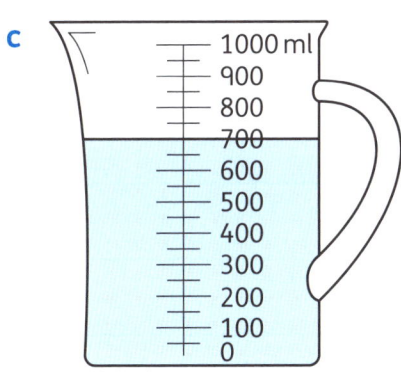

_____ ml

_____ ℓ

d

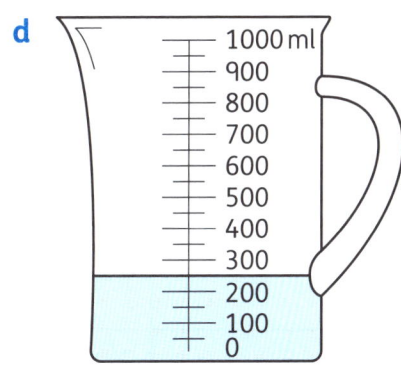

_____ ml

_____ ℓ

e

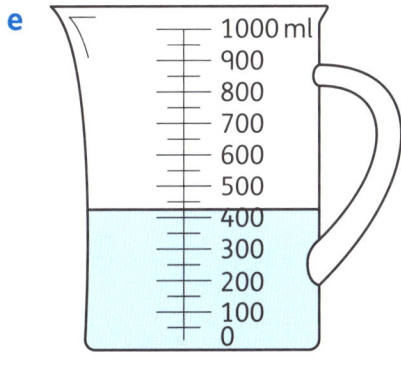

_____ ml

_____ ℓ

f

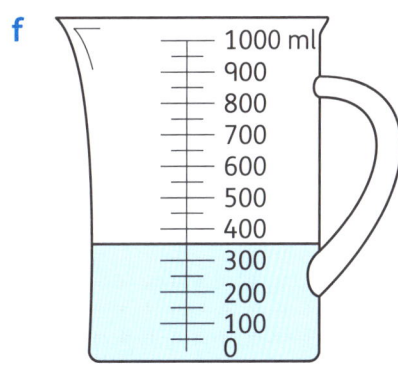
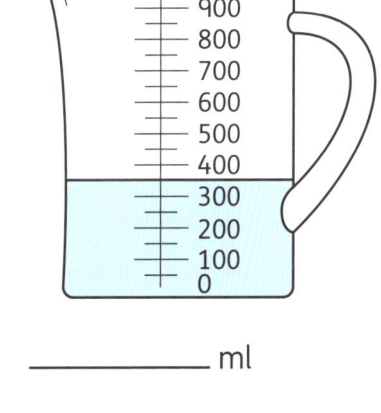

_____ ml

_____ ℓ

➡ *Pupil Book page 44*

Estimate and compare mass

1　Estimate the mass of 1 litre of water.

Use your measurement to complete the chart.

Then check your answers by weighing.

You could weigh a 1-litre bottle of water.
Then tip out the water and weigh the bottle.

Weight of water = weight of bottle of water − weight of empty bottle.

Water	Estimate	Mass
1 litre		
$\frac{1}{2}$ litre		
100 ml		
2 litres		

2　Estimate the mass of each item. Then estimate the masses of some small items in your classroom. Write these in the table.

Check your estimates by weighing the items.

Item	Estimate	Mass
a school bag full of books		
two maths books		
a pair of shoes		
a full lunch box		
an empty lunch box		

3　Circle the best estimate for the mass of each object.

a　a cat　　　　　5 grams　　　　5 kilograms　　　500 grams

b　a beetle　　　2 grams　　　　200 grams　　　2 kilograms

c　a mobile phone　2 grams　　　2 kilograms　　　200 grams

d　an orange　　　100 kilograms　100 grams　　　10 grams

➡ *Pupil Book page 46*

Read measuring scales

1 Draw in the needle to show each mass. The first one has been done for you.

a 3 kg

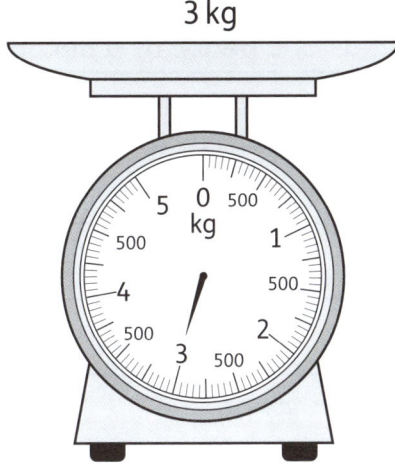

b 2½ kg

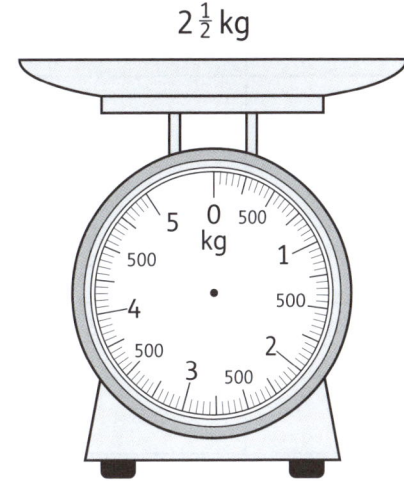

c 1.25 kg

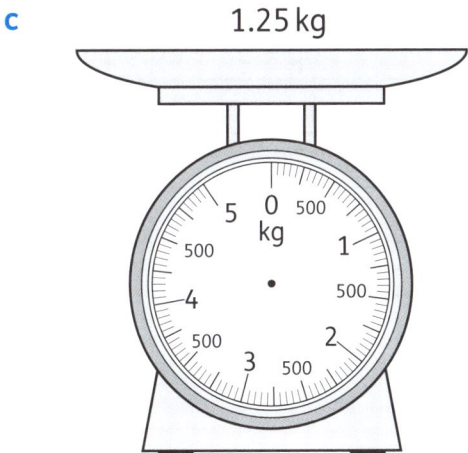

d 3.5 kg

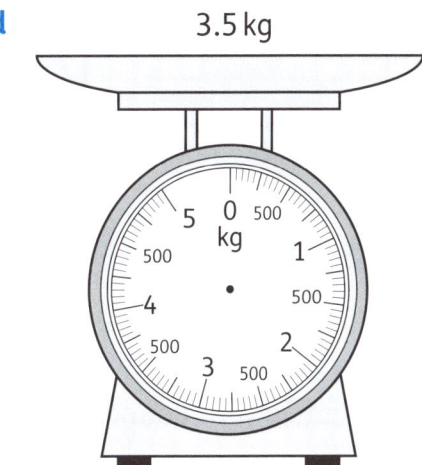

e 4 kg 250 g

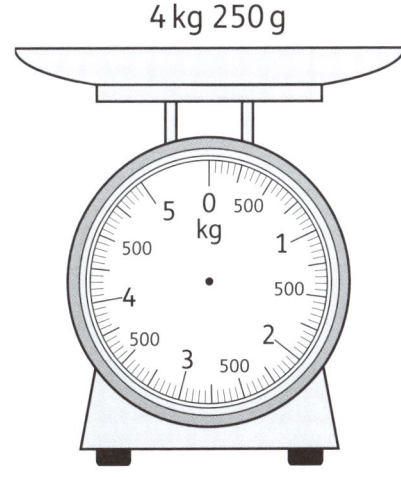

f 2 kg 750 g

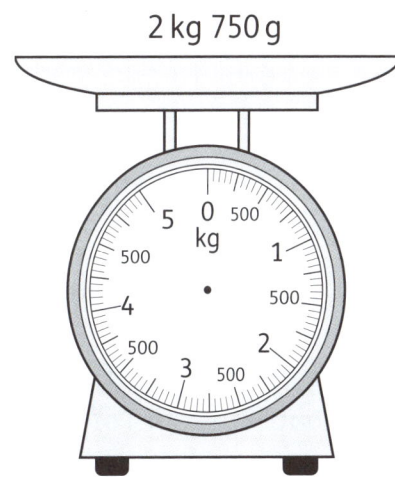

➡ *Pupil Book page 47*

Work with money

Write the amounts in the 3rd column as decimals.

1. Fill in the missing amounts of money to complete the table.

Pounds (£)	Pence (p)	Total in pounds and pence as a decimal
3	40	£3.40
13	5	
12		£12.35
5	99	
7		£7.09
		£14.00

2. Draw lines to match equal amounts of money.

£1.88
£2.99
£4.00
£5.25
£61.50

50p + 50p + 50p + £1 + 20p + 20p + 5p + 2p + 2p
20p + 20p + £2 + 50p + 10p + £1
1p + 2p + 5p + 10p + 20p + £1 + 50p
£50 + £5 + £5 + £1 + 20p + 20p + 10p
£2 + £2 + 50p + 50p + 5p + 10p + 10p

3. a Write these amounts of money in order from most to least money. Assume these amounts are in your country's currency.

| 35.90 | 39.50 | 30.95 | 35.09 | 39.05 | 30.59 |

_____   _____   _____   _____   _____   _____

_____   _____   _____   _____   _____   _____

   b Round each amount to the nearest whole pound/dollar/other. Write the rounded amount below the actual amount.

   c What is the sum of the rounded amounts?

   _____

   d What is double the sum of the rounded amounts?

   _____

▶ Pupil Book page 48

Count and calculate

Count on and back

1 Look for patterns and then count on or back to find the missing numbers on each number line. Write the numbers below each number line.

a
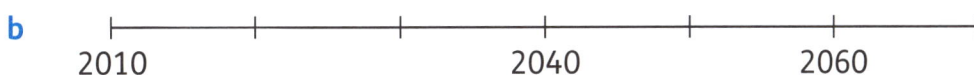

8463   8464   8467

b

2010   2040   2060

c

3810   4010   4110   4210

d

2700   5700   6700

2 Follow the instructions on the track and fill in the missing numbers.

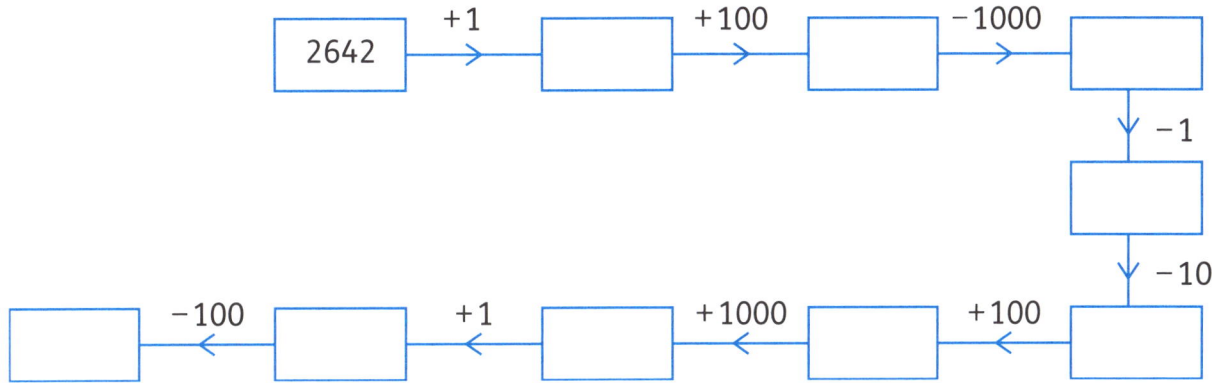

3 Write the correct numbers in the boxes.

a 4000 + 1000 = ☐   **b** ☐ − 1000 = 2430

c 3450 + 1000 = ☐   **d** ☐ − 1000 = 1080

e 8999 + 1000 = ☐   **f** ☐ − 1000 = 999

➡ *Pupil Book page 50*

Count in multiples of 6, 7 and 9

1 Circle the mistake in each sequence. Write the correct number.

a 180   186   192   198   202   Correct number _____

b 300   291   281   273   264   Correct number _____

c 420   427   434   441   447   Correct number _____

d 170   166   160   154   148   Correct number _____

2 Fill in the missing numbers in each sequence of multiples.

a 77   _____   91   98   _____   _____

b 138   132   _____   _____   114   _____

c 189   180   _____   162   _____   _____

d 5058   5067   _____   _____   _____   _____

e 1645   _____   1631   _____   1617   _____

3 Which of these numbers are multiples of 9? Circle them and then write them in order from smallest to greatest.

| 49 | 63 | 72 | 18 | 100 | 99 | 12 | 25 | 27 | 108 | 36 |

_____ _____ _____ _____ _____ _____ _____

4 a Choose a starting number greater than 100. Write it in the circle.

count back in 6s —— —— —— —— ——

count back in 7s —— —— —— —— ——

count back in 9s —— —— —— —— ——

b Compare the numbers that you end on for each counting sequence. What is the difference between them? Can you explain this?

_____

_____

➡ *Pupil Book page 51*

Count in 25s

1 Follow the instructions to complete each chart.

Chart A

⟶ means add 25

↓ means add 1000

Chart B

⟶ means subtract 25

↓ means subtract 1000

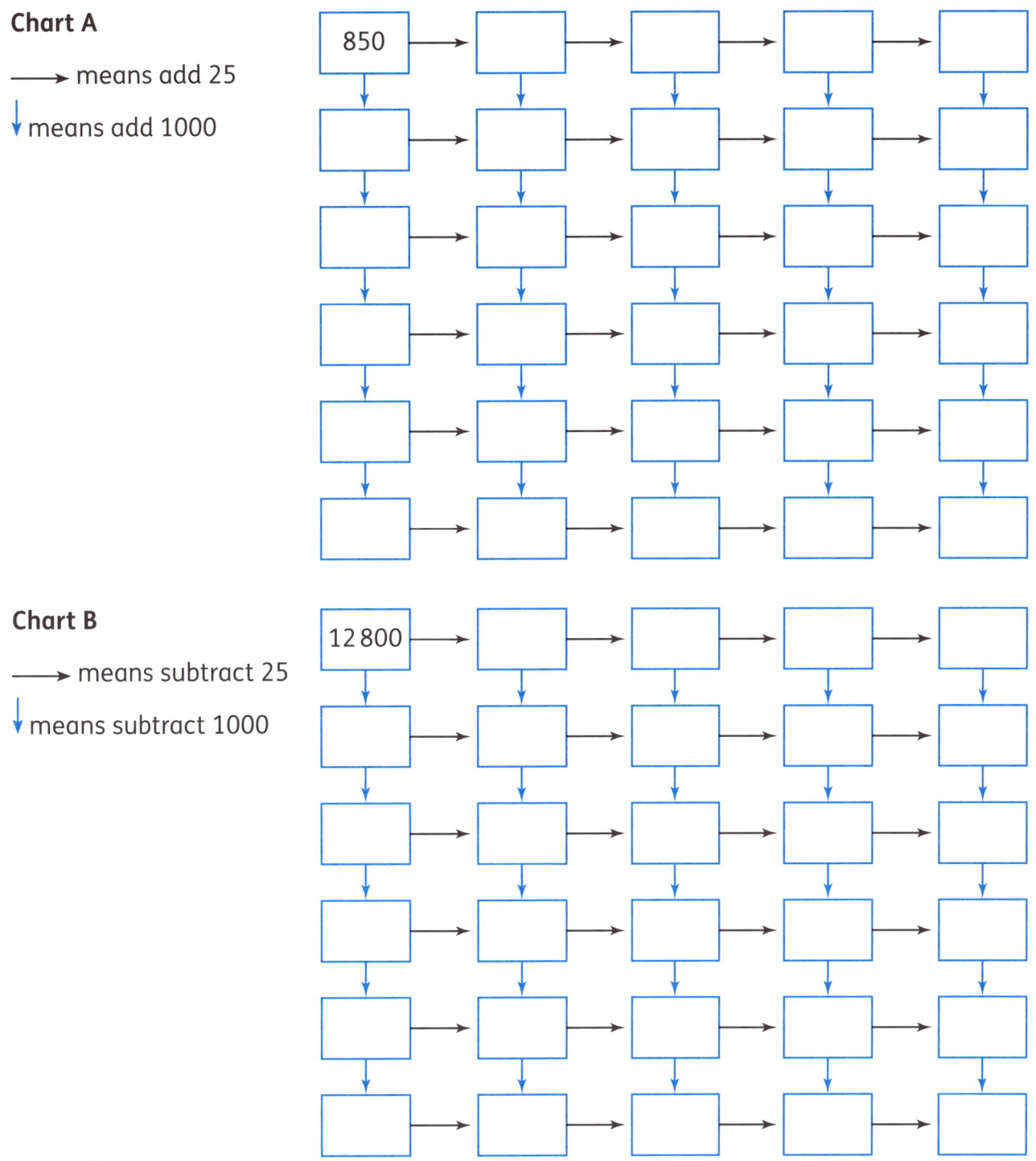

2 When you have completed the charts, look for patterns in the numbers.

Tell your partner about three patterns that you found.

➡ *Pupil Book page 52*

Make 100

1 Sonja cut a 100-cm long piece of ribbon into two pieces. The length of one piece is 45 cm. Write the length of the other piece.

45 cm	

2 Complete the additions. Then write two subtraction facts to match each addition fact. The first one has been done for you.

a 21 + **79** = 100   100 − 21 = 79   100 − 79 = 21

b 25 + ☐ = 100   _____   _____

c 42 + ☐ = 100   _____   _____

d 51 + ☐ = 100   _____   _____

e 64 + ☐ = 100   _____   _____

f ☐ + 90 = 100   _____   _____

g ☐ + 13 = 100   _____   _____

h ☐ + 48 = 100   _____   _____

i ☐ + 77 = 100   _____   _____

3 Look at each set of coins. Write how many more pence each set needs to make £1.

a

20p 5p

b
5p 5p 1p 1p 1p

c
5p 2p 50p 2p

d
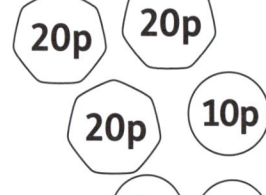
20p 20p 20p 10p 1p 1p

➡ *Pupil Book page 54*

Mental strategies for adding

1 Calculate mentally to find the missing numbers in each diagram. The first one has been done for you.

a
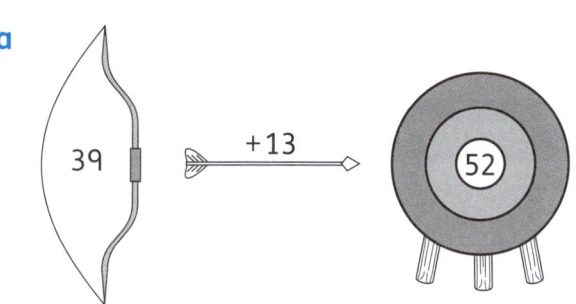
39 +13 52

b
17 +24 ◯

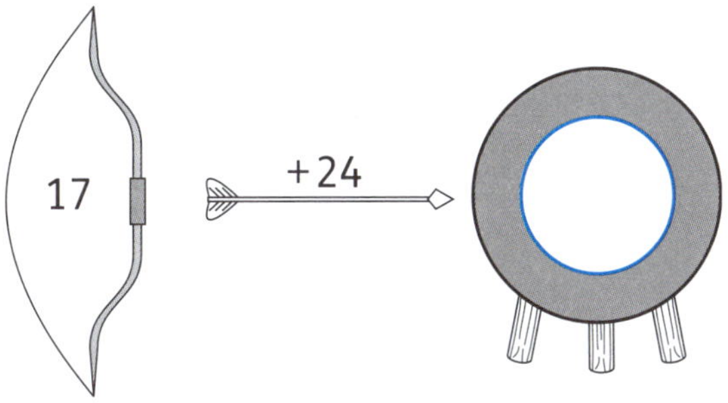

c
☐ +36 62

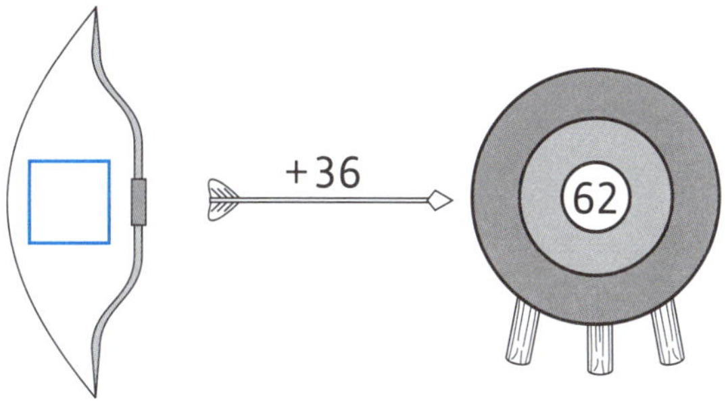

d
37 +48 ◯

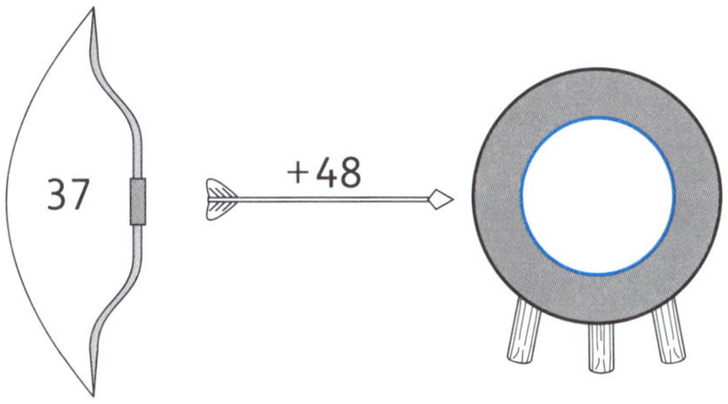

e
48 + ☐ 104

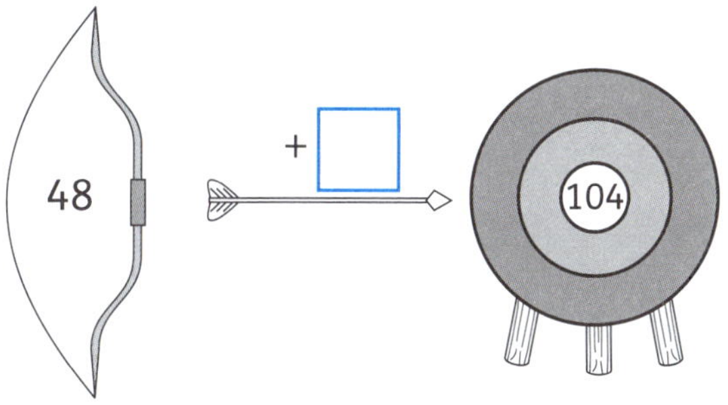

f
☐ +32 91

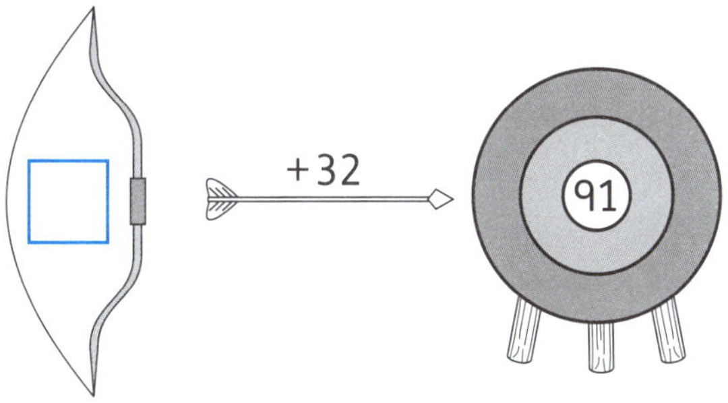

g
63 +19 ◯

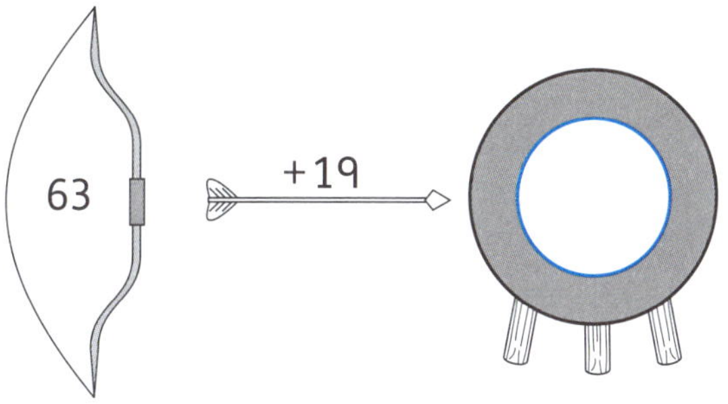

h
☐ +27 102

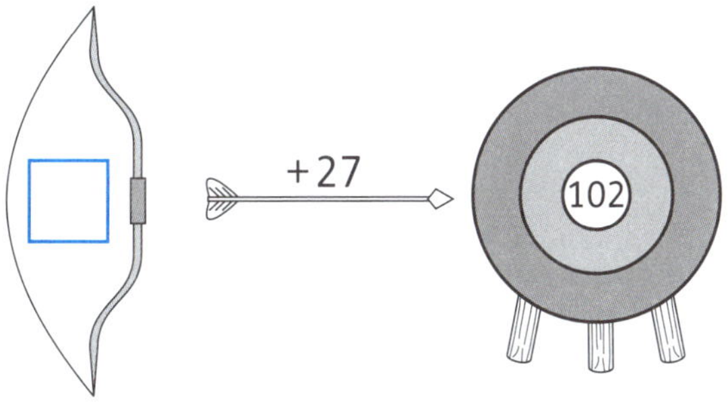

i
84 + ☐ 117

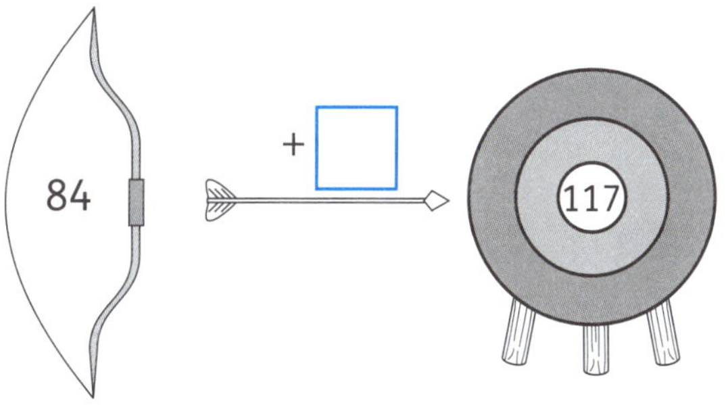

j
92 +19 ◯

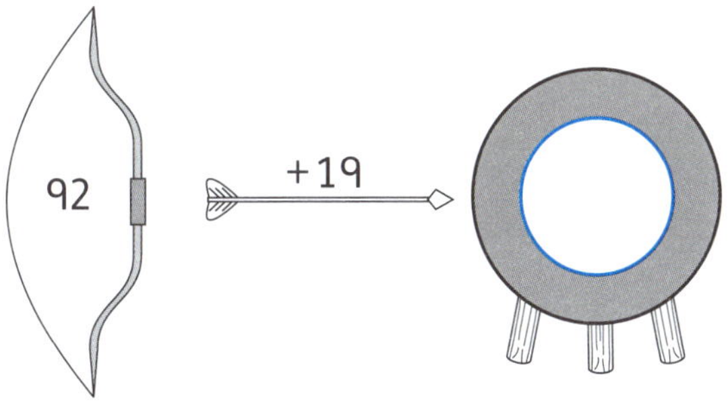

k
☐ +39 80

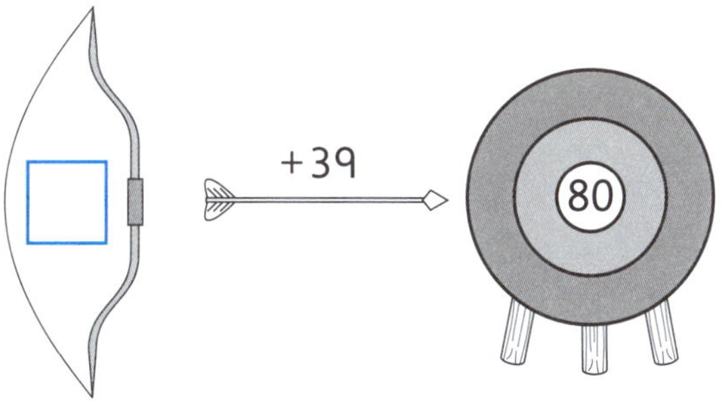

➡ *Pupil Book page 57*

UNIT 8

Symmetry

Symmetrical shapes

1 The dashed lines on each shape are lines of symmetry.
Draw the other half of each shape.

> Look carefully and draw one line at a time.

a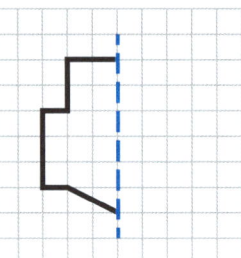

b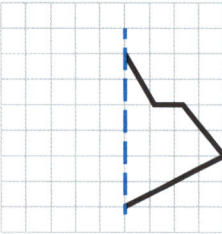

c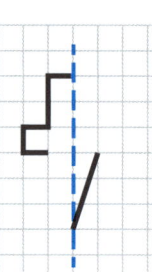

d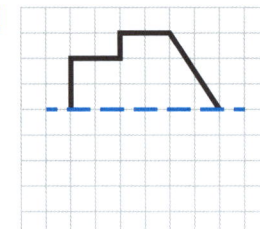

e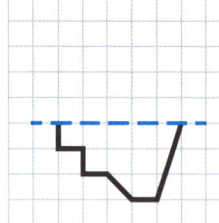

f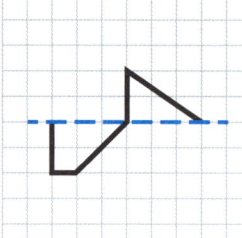

2 Complete these symmetrical shapes.

a

b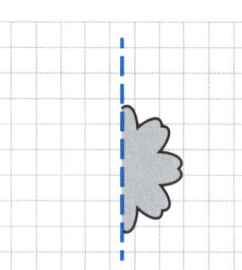

c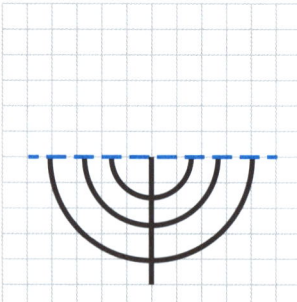

 Problem solving

Both dashed lines are lines of symmetry on these grids.

3 Complete this shape.

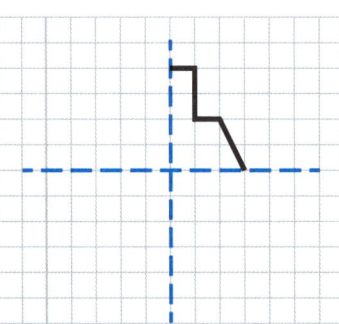

4 Draw your own symmetrical shape on this grid.

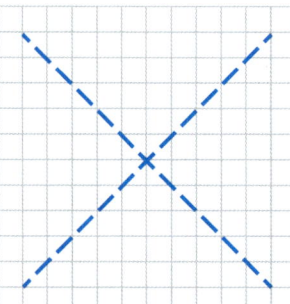

➡ *Pupil Book page 62*

Investigating lines of symmetry (1)

> A line of symmetry divides a shape into two identical parts.

1 Write the name of each shape.

Draw at least one line of symmetry on each shape.

a

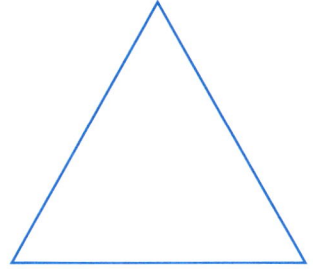

_____

b

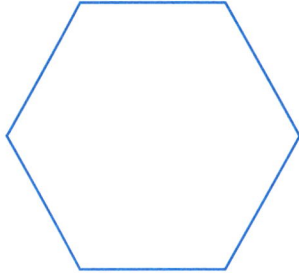

_____

c

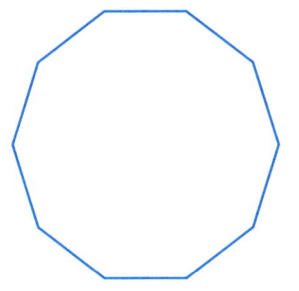

_____

d

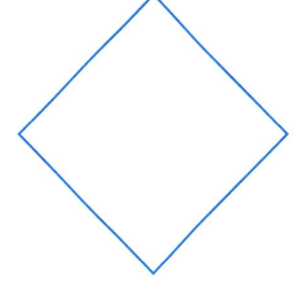

_____

e

_____

f

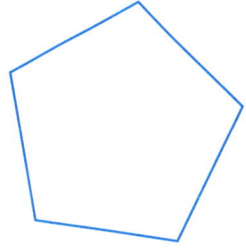

_____

g

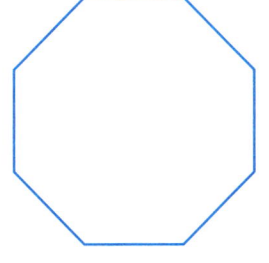

_____

h

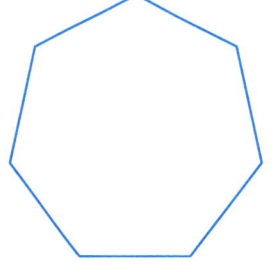

_____

➡ *Pupil Book page 63*

Investigating lines of symmetry (2)

1 Draw the lines of symmetry on each shape.
Use a different colour for each line of symmetry.

Write the name of each shape you know.

Look carefully and draw one line at a time.

a
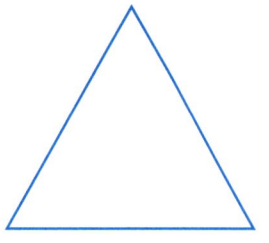

_____

b

_____

c
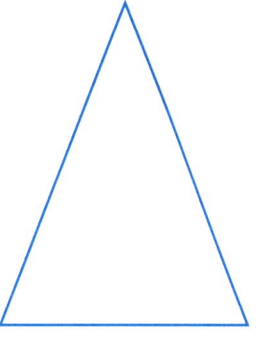

_____

d

_____

e
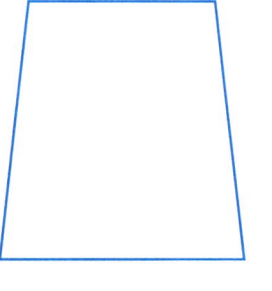

_____

f
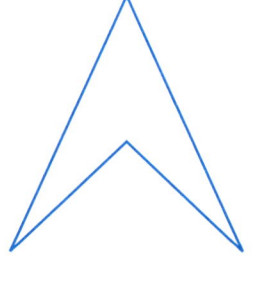

_____

g
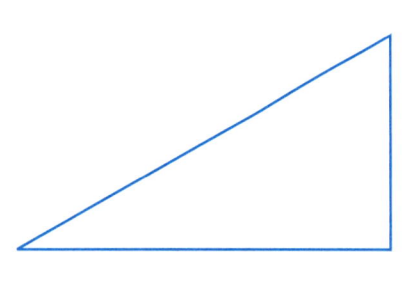

_____

h
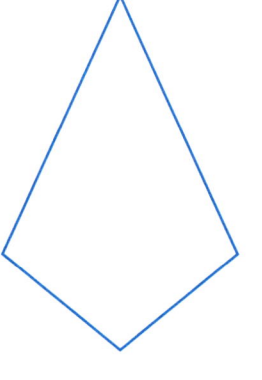

_____

➡ *Pupil Book page 63*

Symmetry around us

You can use dotted paper to make a symmetrical design.

1 Draw the lines of symmetry on this rangoli pattern and then colour it. Your colouring should be symmetrical.

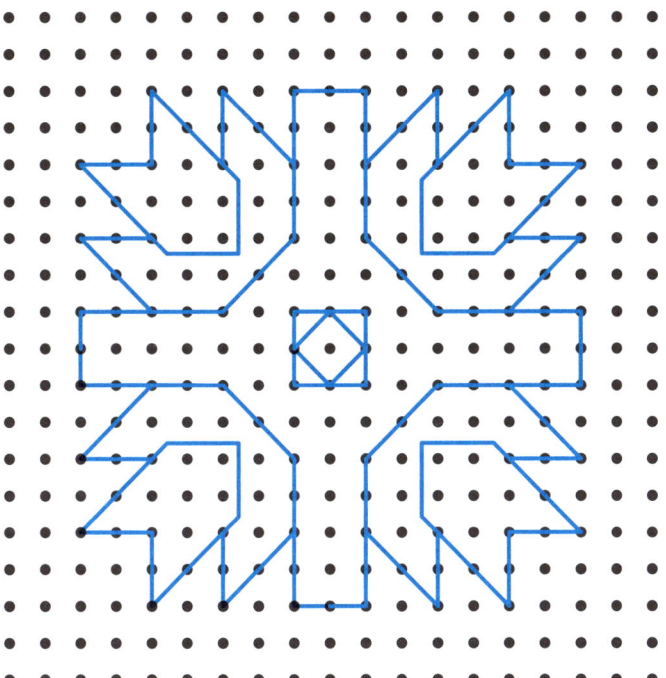

2 Find a symmetrical design that you like. It could be a traditional design, a logo on a T-shirt, or a pattern on a fabric or tile.

Draw the design on this grid.

➡ *Pupil Book page 64*

Data and charts

Frequency tables

1. Work with a partner. You will need 6 cards numbered from 1 to 6.

   Put them into a bag. Pick a number (without looking) and record it in the frequency table. Then put it back in the bag. Take turns.

   Do this 30 times.

   > The frequency is the total of the tallies.

Number on card	Tally	Frequency
1		
2		
3		
4		
5		
6		

2. In the frequency table, record the number of times the vowels 'a', 'e', 'i', 'o' and 'u' appear in the paragraph below.

   > Mario did a survey to find out which colours the pupils in a Year 4 group liked best. The survey form gave them four choices: red, blue, white and yellow. More pupils chose blue than red and green combined. Twenty-three pupils were surveyed and only three chose white as their favourite.

   > Add headings to the columns.

   a Which vowel occurs most frequently? _____

   b How many more times does 'e' occur than 'u'? _____

   c How many vowels did you count altogether? _____

➡ *Pupil Book page 66*

Bar charts

1 The frequency table shows the results of spinning a spinner.

Results	Tally	Frequency
blue	̶H̶H̶T̶ ̶H̶H̶T̶ ̶H̶H̶T̶ ̶H̶H̶T̶ II	
green	̶H̶H̶T̶ ̶H̶H̶T̶ II	
yellow	̶H̶H̶T̶ II	
red	̶H̶H̶T̶ IIII	
	Total number of spins:	

a Complete the chart by filling in the frequency of each colour and the total.

b Draw a bar chart to show the results.

> Remember to give the graph a heading and to label the scales.

Problem solving

2 This is the spinner that was used.

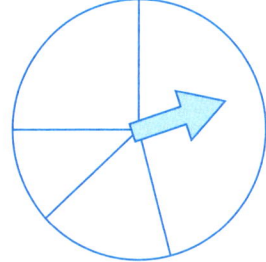

a What colour should each section be? Colour the spinner to show your answer.

b How did you decide which colour to use for each section?

_____

_____

➡ *Pupil Book page 67*

More bar charts

You will need a small counter to complete this activity.

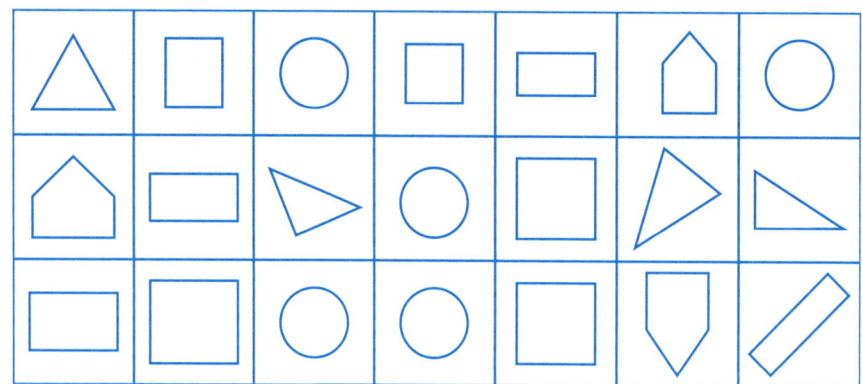

1 Toss the counter onto the game board. If it does not land on a shape, push it so it does.

Use the frequency table to record the shapes you land on. Repeat this for 20 tosses.

Shape	Tally	Frequency
triangle		
square		
rectangle		
circle		
pentagon		

2 Write a title and label to complete this bar chart to show your results.

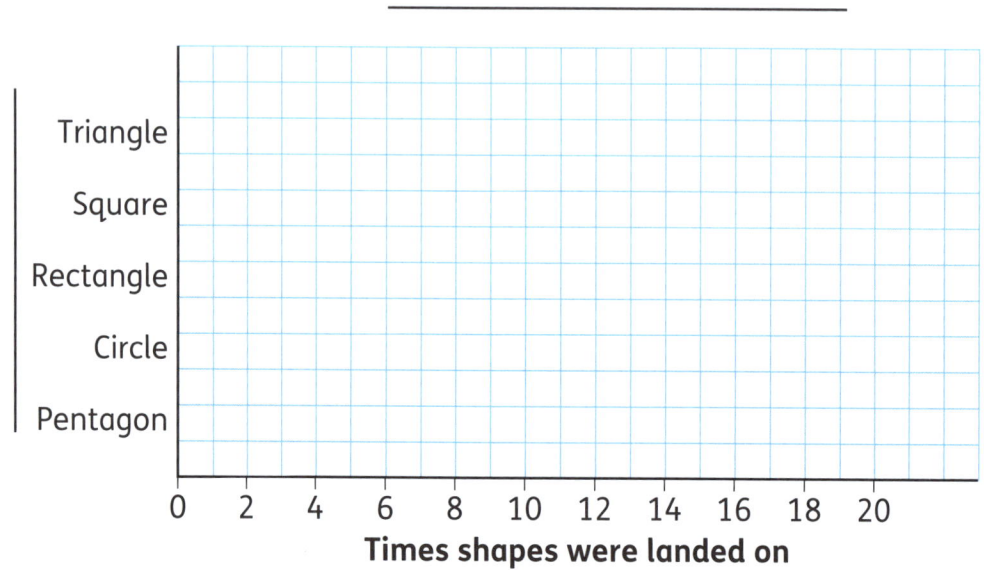

3 Compare bar charts with a partner.

    a How are they similar? _____

    b How are they different? _____

➡ *Pupil Book page 69*

Pictograms

1 The table shows the number of tourists who took a bus tour on each day of the week.

Day	Number of tourists	Frequency
Monday	ⲎⲎⲎ ⲎⲎⲎ ⲎⲎⲎ II	
Tuesday	ⲎⲎⲎ ⲎⲎⲎ III	
Wednesday	ⲎⲎⲎ ⲎⲎⲎ ⲎⲎⲎ ⲎⲎⲎ I	
Thursday	ⲎⲎⲎ ⲎⲎⲎ ⲎⲎⲎ	
Friday	ⲎⲎⲎ ⲎⲎⲎ ⲎⲎⲎ ⲎⲎⲎ I	
Saturday	ⲎⲎⲎ ⲎⲎⲎ ⲎⲎⲎ ⲎⲎⲎ ⲎⲎⲎ ⲎⲎⲎ II	
Sunday	ⲎⲎⲎ ⲎⲎⲎ ⲎⲎⲎ ⲎⲎⲎ ⲎⲎⲎ ⲎⲎⲎ IIII	

a Complete the table by filling in the frequency for each day.

b Draw a pictogram to show this data. Add a title.

_____

Monday	
Tuesday	
Wednesday	
Thursday	
Friday	
Saturday	
Sunday	

Key: ⊗ = 5 people

c Write two sentences about what your pictogram shows.

_____

_____

➡ *Pupil Book page 71*

Venn diagrams

1 Draw at least two 2D shapes in each section of these Venn diagrams. The first one has been started for you.

a

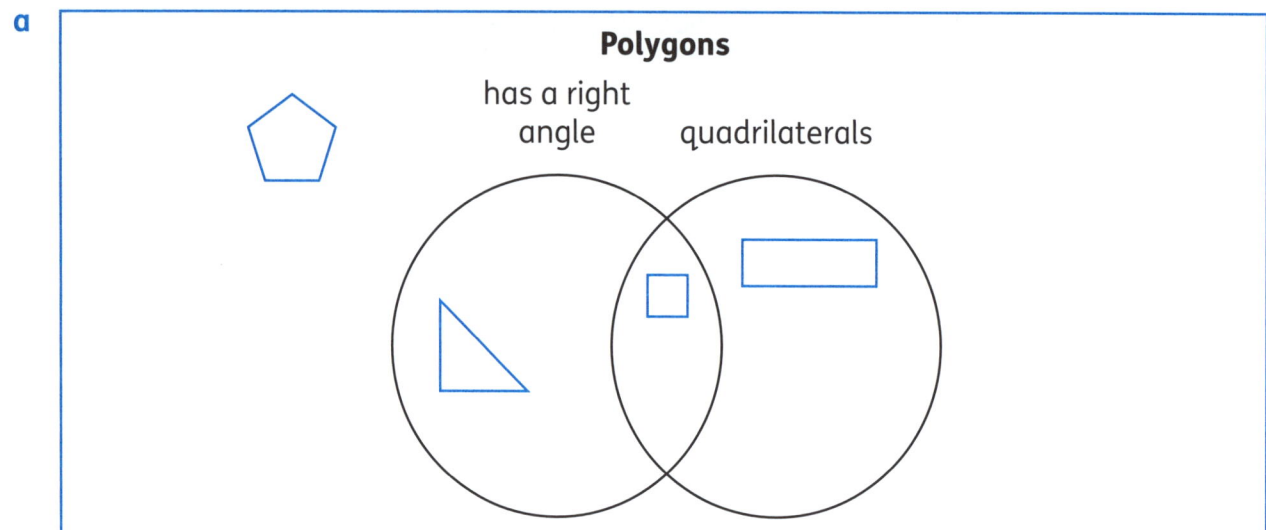

b

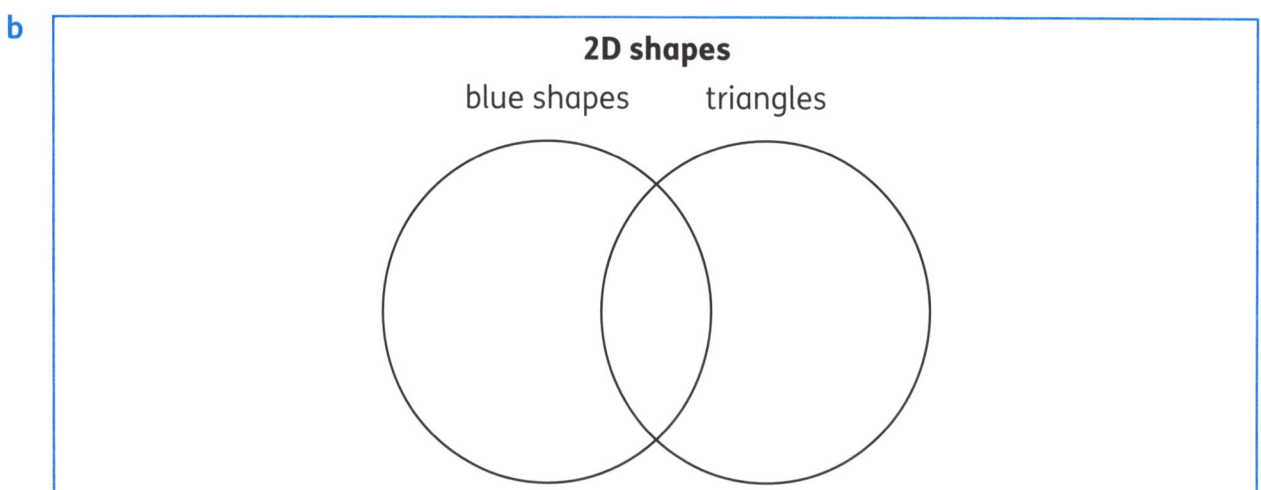

2 Write all the numbers from 1 to 30 in the correct places on this Venn diagram.

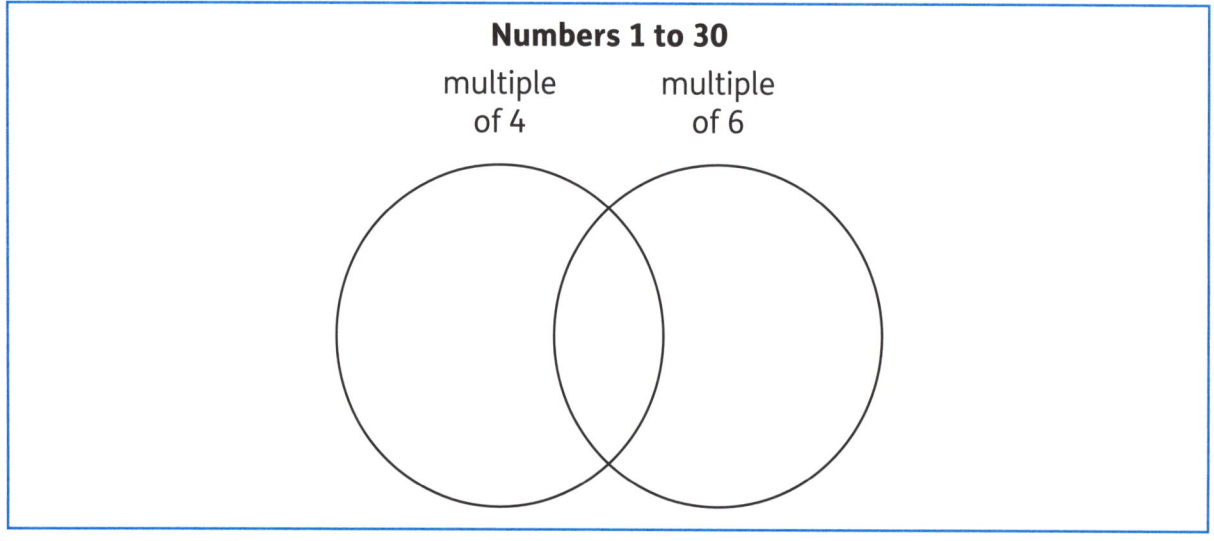

➡ *Pupil Book page 72*

Solving problems using charts

1 This incomplete bar chart shows how fast different animals can run.

An ostrich can run at 72 km per hour and a giraffe can run at 60 km per hour. Use this data to complete the graph.

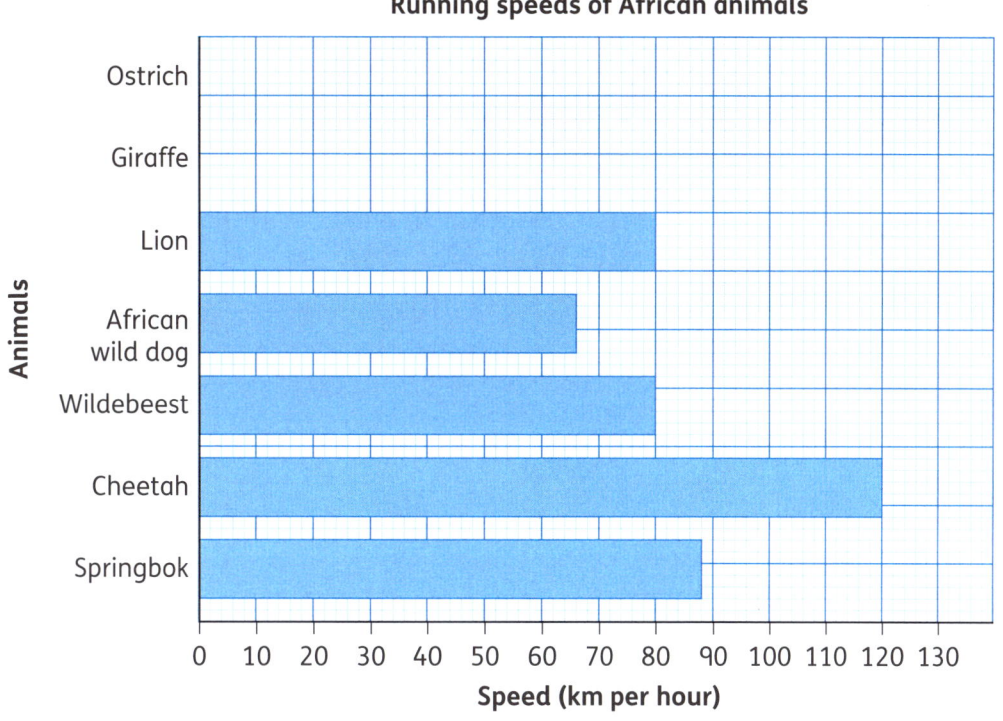

Running speeds of African animals

2 Answer these questions about the bar chart.

a What is the fastest speed that a springbok can run?

_____

b Which animal has the fastest running speed?

_____

c Which animals can run at 80 km per hour?

_____

d How much faster can a cheetah run than a lion?

_____

e What is the running speed of the slowest animal shown on the graph?

_____

f Usain Bolt set an Olympic Record (2009) of 9.58 seconds for the 100 m sprint. This is 37.58 km per hour. How much faster than this can a cheetah run?

_____

➡ *Pupil Book page 73*

Function machines

1 Fill in the missing numbers in each function machine.

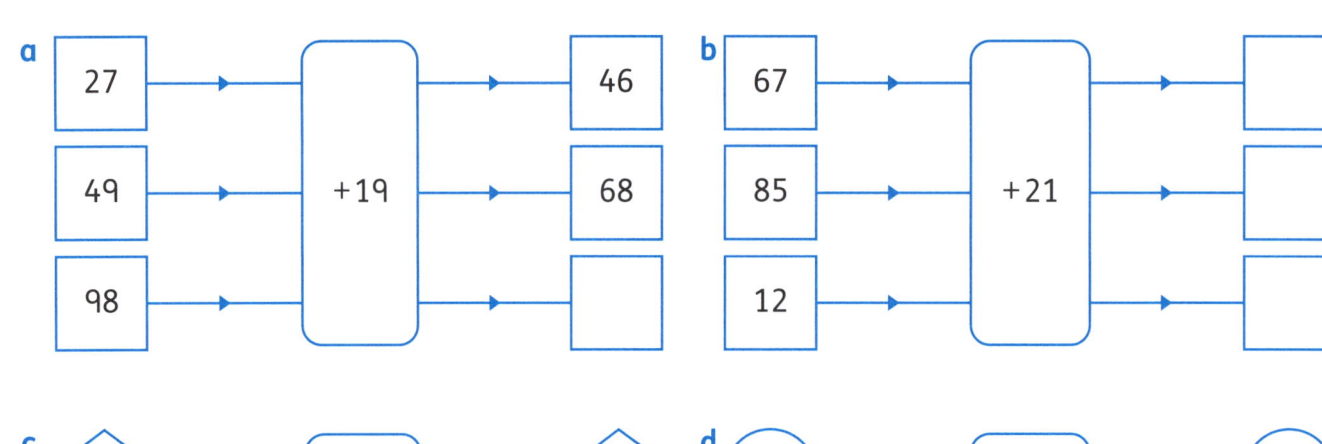

a
27	→	+19	→	46
49	→		→	68
98	→		→	

b
67	→	+21	→	
85	→		→	
12	→		→	

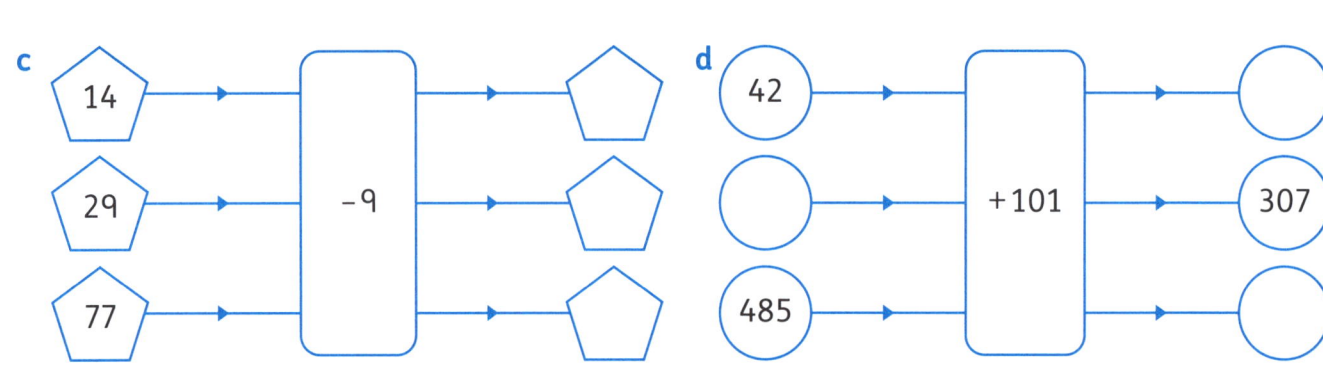

c
14	→	−9	→	
29	→		→	
77	→		→	

d
42	→	+101	→	
	→		→	307
485	→		→	

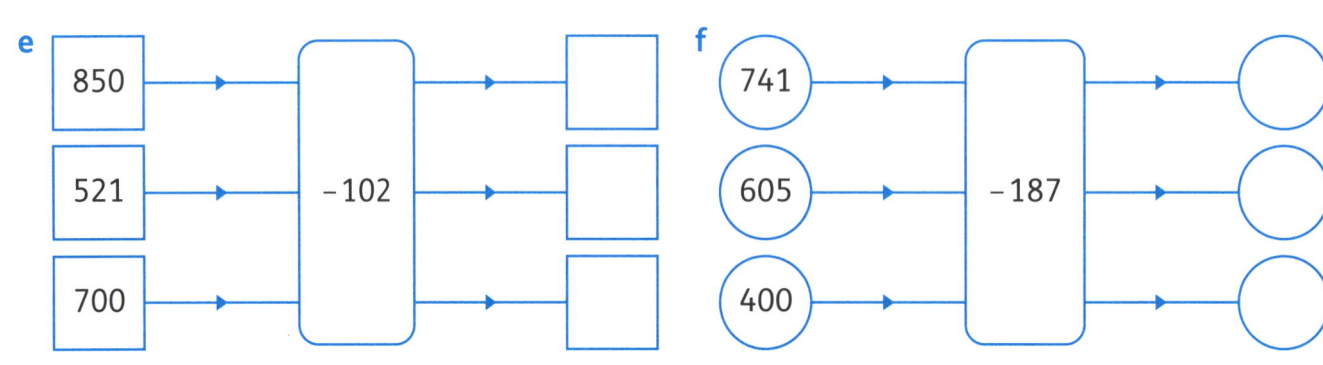

e
850	→	−102	→	
521	→		→	
700	→		→	

f
741	→	−187	→	
605	→		→	
400	→		→	

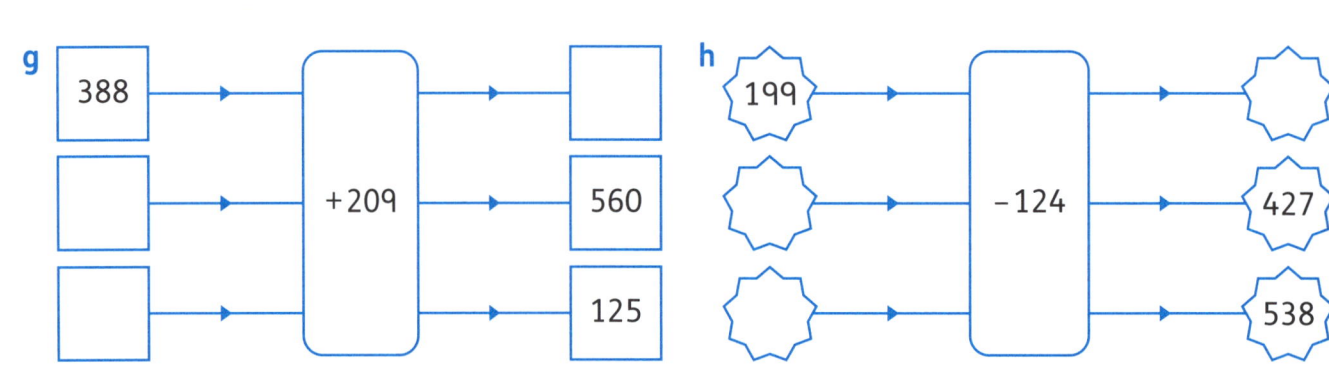

g
388	→	+209	→	
	→		→	560
	→		→	125

h
199	→	−124	→	
	→		→	427
	→		→	538

➡ *Pupil Book page 75*

Number puzzles

1 Put numbers in the empty boxes to complete the puzzles.

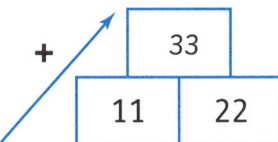

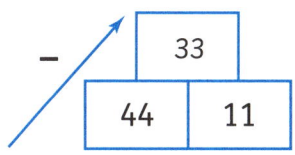

Look out for the signs.

Add the numbers in the two boxes and write the answer in the top box.

Find the difference between the two numbers and write the answer in the top box.

a

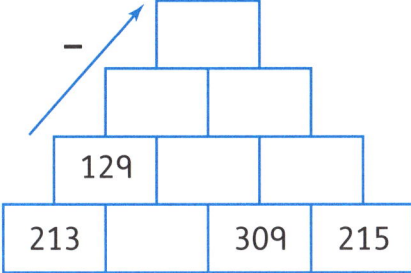

b

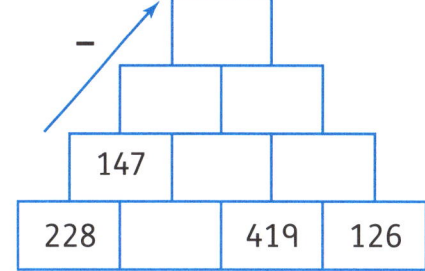

c

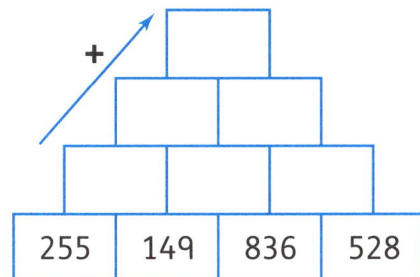

d

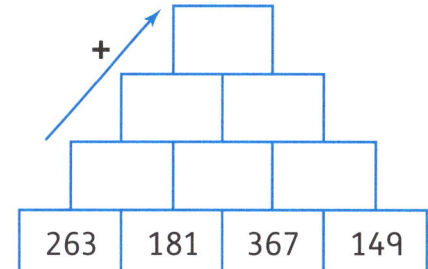

e

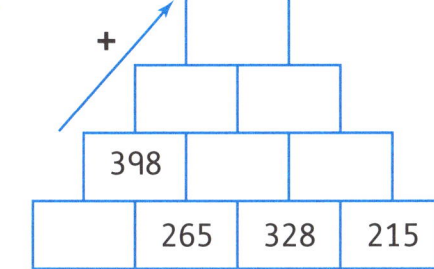

2 Use this shape to make your own puzzle.

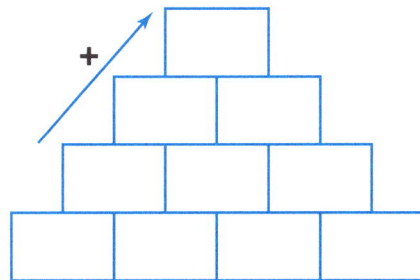

➡ *Pupil Book page 78*

1 Complete the target diagrams. The two numbers in a section must add up to the number in the centre. One section has been done for you.

a

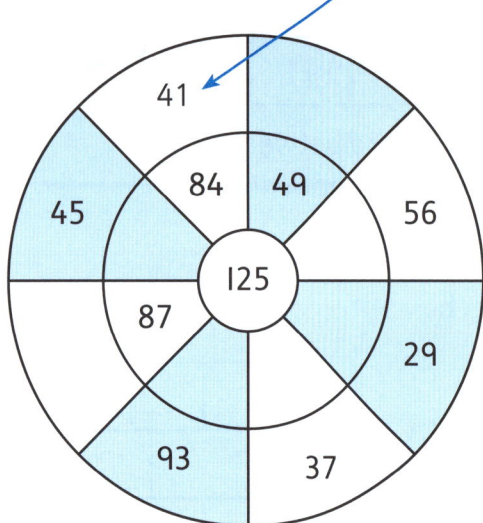

b

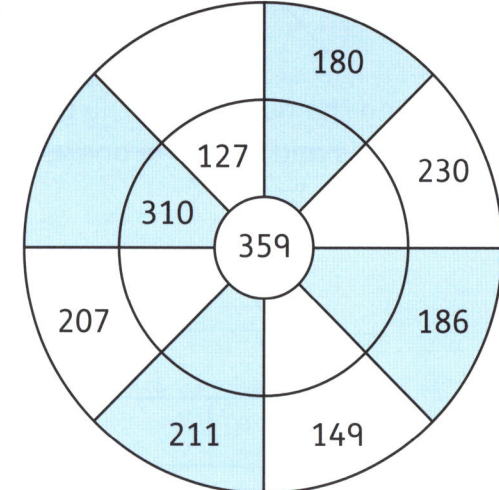

c

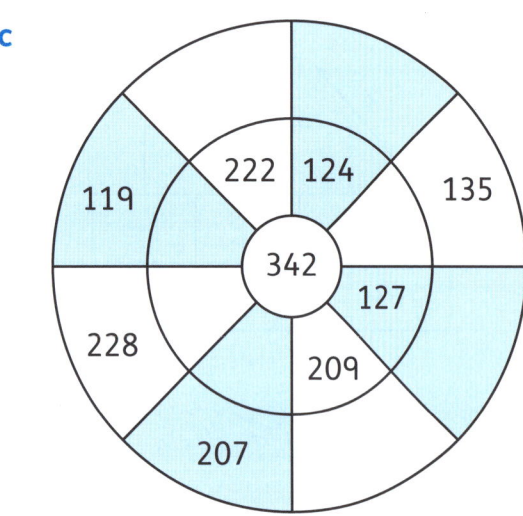

d

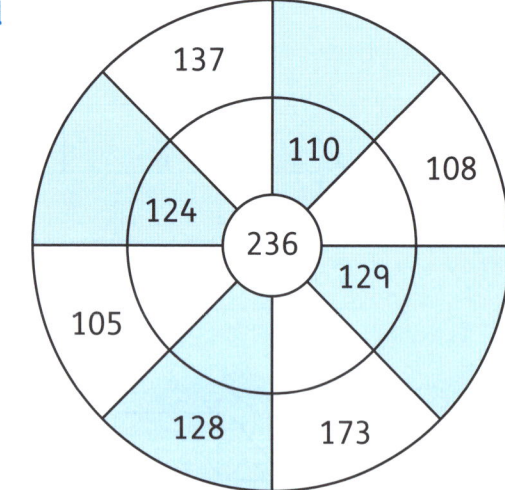

e

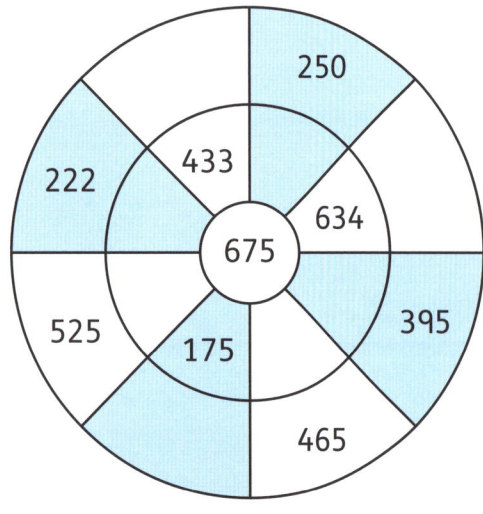

f

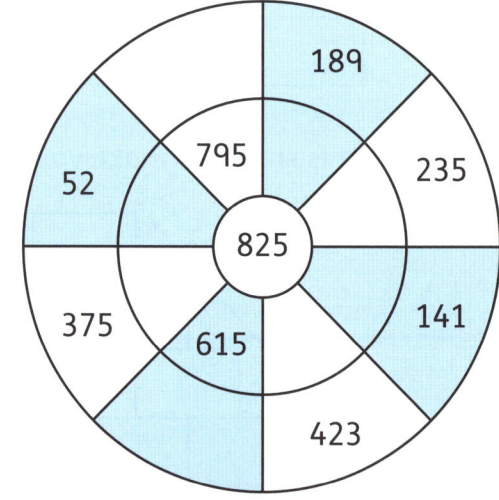

➡ *Pupil Book page 79*

Angles and triangles

Compare and order angles (1)

1. Use three different colours. Colour the right angles in one colour, the acute angles in another and the obtuse angles in the third colour.

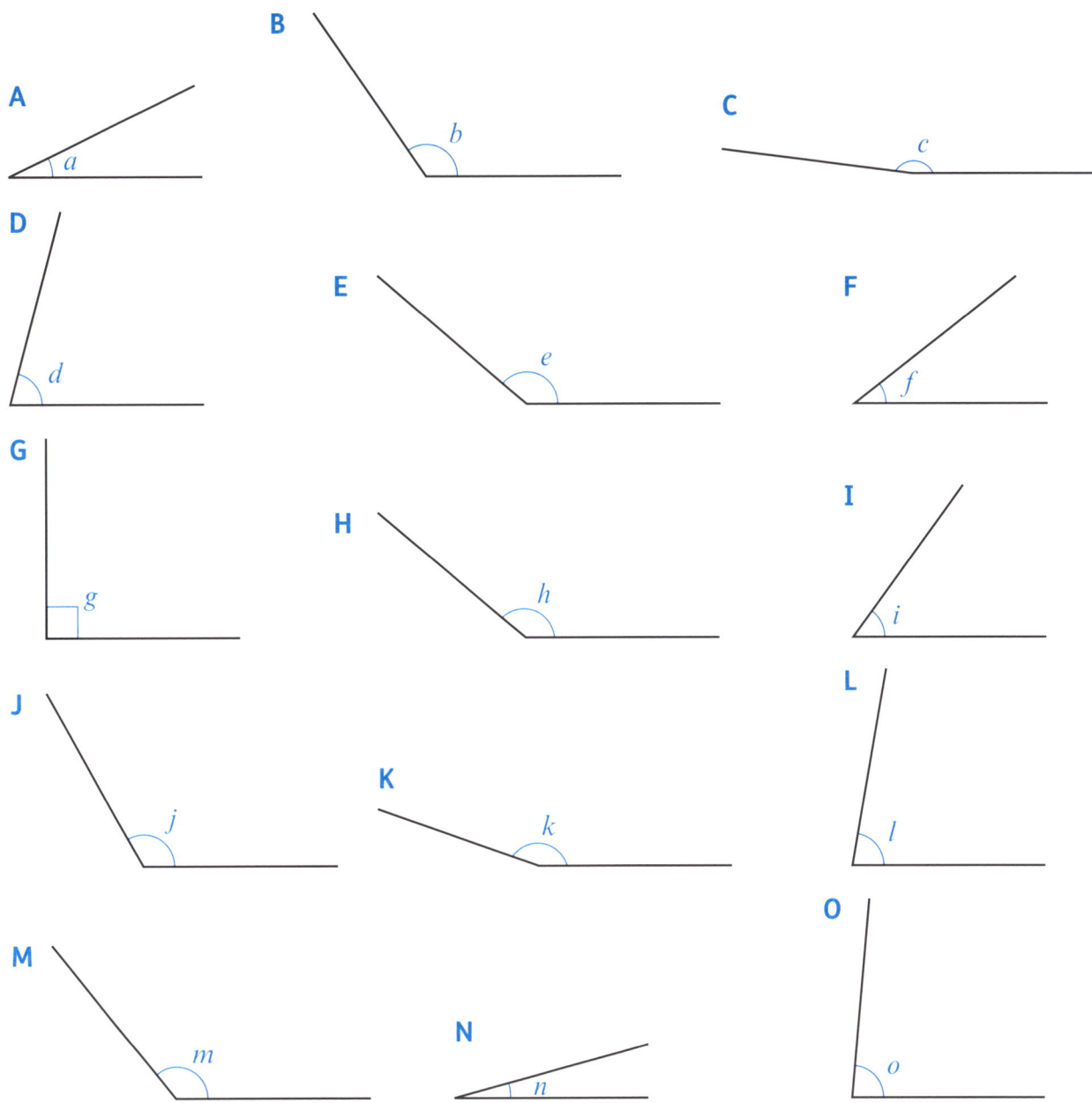

2. Use this table to sort the angles you coloured in question 1. Write the letters in the correct columns.

Acute angles	Right angles	Obtuse angles

➡ *Pupil Book page 83*

Compare and order angles (2)

1 Sonja wrote her name on squared paper using only straight lines. She marked all the right, acute and obtuse angles inside the letters in different ways.

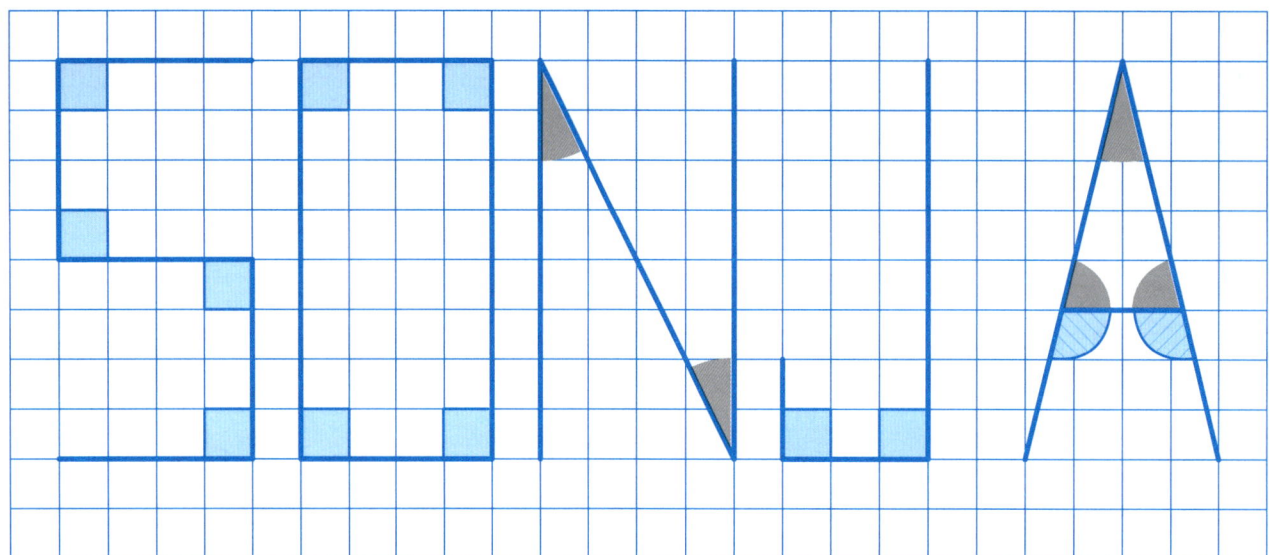

a Write your own name in the same way. Mark the different angles.

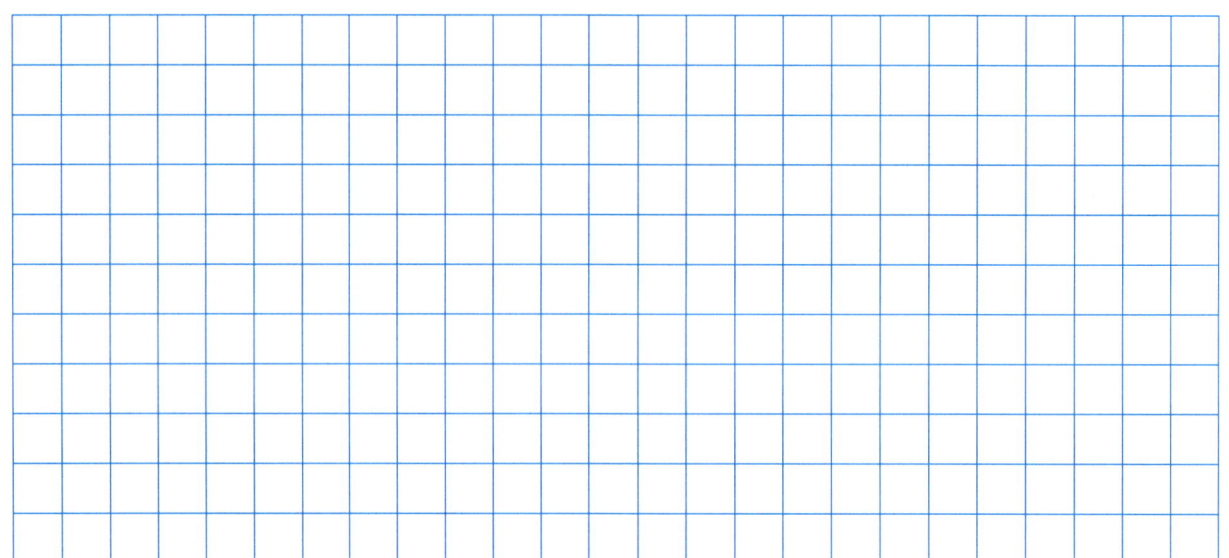

b Complete this chart to count the number of each type of angle in your name.

Type of angles	Tally	Frequency
right angles		
acute angles		
obtuse angles		

➡ *Pupil Book page 83*

Classify triangles

1 Colour the odd triangle out in each set. You may need to measure the sides and angles to help you decide which one is the odd one out.

Write a sentence saying why the coloured one does not fit with the others in the set.

a

_____

b

_____

c

_____

d

_____

▶ *Pupil Book page 85*

Multiplication and division facts

Learn your facts

1 You can use the tables square to help you practise multiplication.

 a Complete the tables square.

 b Ask someone to test you, using the tables square to find multiplication questions.

 You can also use the square to check the answers.

×	1	2	3	4	5	6	7	8	9	10	11	12
1	1	2	3	4	5	6	7	8	9	10	11	12
2	2	4	6	8	10	12	14	16	18	20	22	24
3	3	6	9	12	15			27		30	33	36
4	4	8	12	16	20			32		40	44	48
5	5	10	15	20	25	30				50	55	60
6	6	12			30		42			60	66	72
7	7	14			35		49	56		70	77	84
8	8	16		32						80	88	96
9	9	18							81	90	99	108
10	10	20	30	40	50	60	70	80	90	100	110	120
11	11	22	33	44	55	66	77	88	99	110	121	132
12	12	24	36	48	60	72	84	96	108	120	132	144

2 Fill in the missing numbers.

 a $\boxed{} \times 6 = 42$

 b $7 \times \boxed{} = 35$

 c $5 \times \boxed{} = 20$

 d $9 \times \boxed{} = 27$

 e $\boxed{} \times 8 = 56$

 f $\boxed{} \times 12 = 36$

▶ *Pupil Book page 87*

Times table strips

This is a times table strip showing multiples of 4.

×	1	2	3	4	5	6	7	8	9	10	11	12
4	4	8	12	16	20	24	28	32	36	40	44	48

---

1 Complete each times table strip by filling in the missing numbers.

a

×	1	2	3	4	5	6	7	8	9	10	11	12
5					25					50		

b

×	1	2	3	4	5	6	7	8	9	10	11	12
6	6	12	18									

c

×	1	2	3	4	5	6	7	8	9	10	11	12
7					35					70		

d

×	1	2	3	4	5	6	7	8	9	10	11	12
8						48	56					

e

×	1	2	3	4	5	6	7	8	9	10	11	12
9			27	36								

f

×	1	2	3	4	5	6	7	8	9	10	11	12
10	10	20										

g

×	1	2	3	4	5	6	7	8	9	10	11	12
11	11	22	33									

h

×	1	2	3	4	5	6	7	8	9	10	11	12
12									108	120	132	144

➡ *Pupil Book page 90*

Division facts

We can make cards to show the numbers in a fact family.

This card has 7, 8 and 56 on it because 7 × 8 = 56.

7	56	8

> Remember that a multiplication and divison fact family is all the facts for three numbers.

These cards are similar, but one number has been left out on each.

---

1 Write the missing numbers.

a
5		9

b
4	36	

c
	27	9

d
7	35	

e
9		7

f
8	56	

g
	28	4

h
8	24	

i
6	36	

j
7		3

k
6	42	

l
	40	8

m
8	72	

n
8		6

o
6		3

p
8		8

q
6		9

r
	24	6

2 Complete these tables of division facts.

a
	÷ 2	
	÷ 3	
24	÷ 4	
	÷ 6	
	÷ 8	

b
	÷ 2	
	÷ 3	
30	÷ 5	
	÷ 6	
	÷ 10	

c
	÷ 2	
	÷ 4	
40	÷ 5	
	÷ 8	
	÷ 10	

d
	÷ 2	
	÷ 4	
48	÷ 6	
	÷ 8	
	÷ 12	

➡ *Pupil Book page 92*

Use properties of multiplication and division

1. Choose three factors from the box each time and find their product.
   Try to find 9 different products.

0	1	2	3	4	5

☐ × ☐ × ☐ = ☐   ☐ × ☐ × ☐ = ☐   ☐ × ☐ × ☐ = ☐

☐ × ☐ × ☐ = ☐   ☐ × ☐ × ☐ = ☐   ☐ × ☐ × ☐ = ☐

☐ × ☐ × ☐ = ☐   ☐ × ☐ × ☐ = ☐   ☐ × ☐ × ☐ = ☐

2. Write three different ways of doing each multiplication and write the product.
   The first one has been done for you.

a $2 \times 4 \times 6$ = 8 × 6 = 2 × 24 = 4 × 12 = 48

b $2 \times 6 \times 5$ = ☐ × ☐ = ☐ × ☐ = ☐ × ☐ = ☐

c $10 \times 3 \times 6$ = ☐ × ☐ = ☐ × ☐ = ☐ × ☐ = ☐

d $9 \times 2 \times 3$ = ☐ × ☐ = ☐ × ☐ = ☐ × ☐ = ☐

e $1 \times 4 \times 7$ = ☐ × ☐ = ☐ × ☐ = ☐ × ☐ = ☐

f $11 \times 2 \times 0$ = ☐ × ☐ = ☐ × ☐ = ☐ × ☐ = ☐

3. Choose three different digits from the cards to complete each multiplication.

a ☐ × ☐ × ☐ = 24

b ☐ × ☐ × ☐ = 72

c ☐ × ☐ × ☐ = 60

d ☐ × ☐ × ☐ = 70

e ☐ × ☐ × ☐ = 30

f ☐ × ☐ × ☐ = 120

2	3	4	5	6	7

➡ *Pupil Book page 95*

Negative numbers

Numbers less than 0

1 Fill in the numbers on each number line.

a

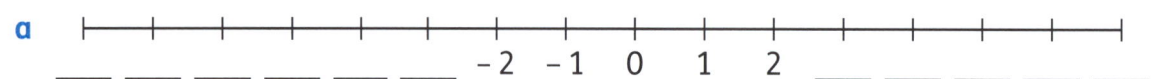

___ ___ ___ ___ ___ ___ –2 –1 0 1 2 ___ ___ ___ ___ ___

b

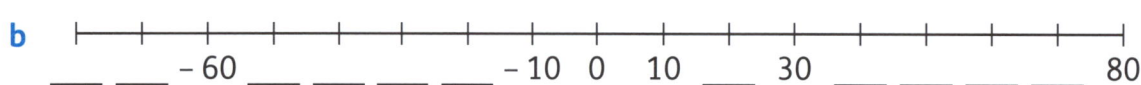

___ ___ – 60 ___ ___ ___ ___ – 10 0 10 ___ 30 ___ ___ ___ ___ 80

c

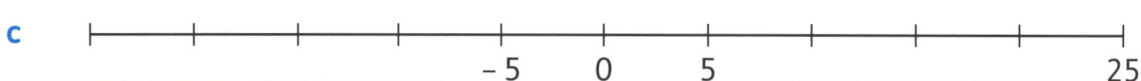

____ ____ ____ ____ – 5 0 5 ____ ____ ____ 25

2 Count back in steps to complete these number patterns. Use the number lines above to help you.

a 30   20   10   0   _____   _____

b 15   10   5   _____   _____   _____

c 12   _____   4   0   _____   _____

d – 10   – 15   _____   _____   _____   _____

e 16   8   _____   – 8   _____   _____

f 120   20   _____   – 180   _____   _____

3 Complete this number chain:

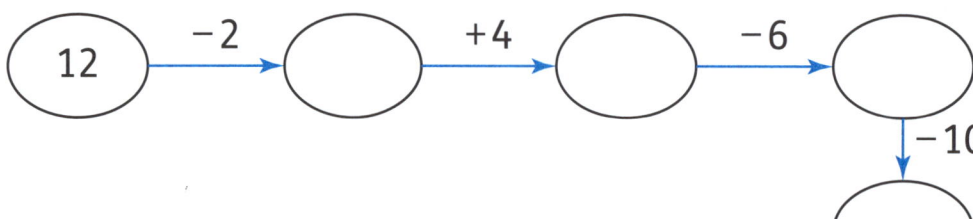

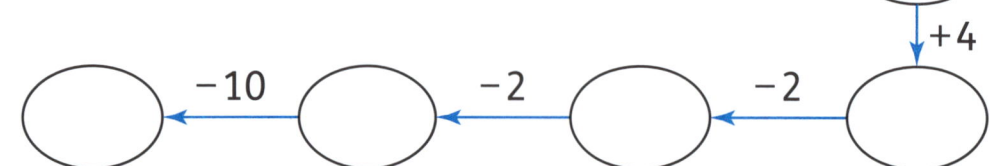

➡ *Pupil Book page 96*

Perimeter and area

Perimeter (1)

1 Measure the length and width of each rectangle. Write the measurements on the diagrams.

 Perimeter is the distance around a closed shape.

2 Calculate the perimeter of each rectangle.

a

Perimeter = _____

b

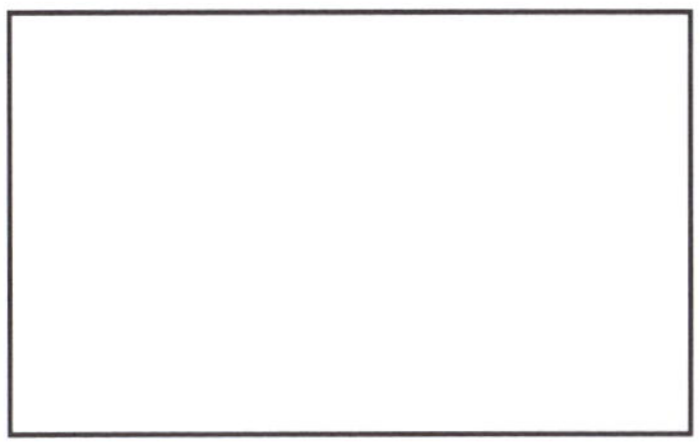

Perimeter = _____

c

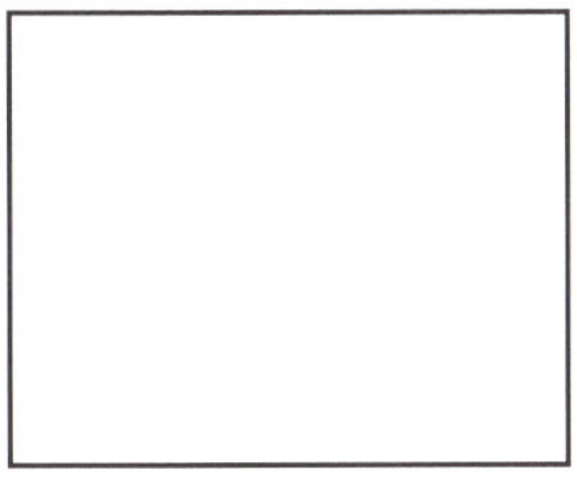

Perimeter = _____

d

Perimeter = _____

e

Perimeter = _____

f

Perimeter = _____

 Remember to write the units.

➡ *Pupil Book page 99*

Perimeter (2)

Remember, perimeter is the distance around a closed shape.

1 Draw three different rectangles with a perimeter of 24 cm.

➡ *Pupil Book page 99*

1 Each diagram shows part of a rectangle. You are given the perimeter (P) of each rectangle. Complete each diagram.

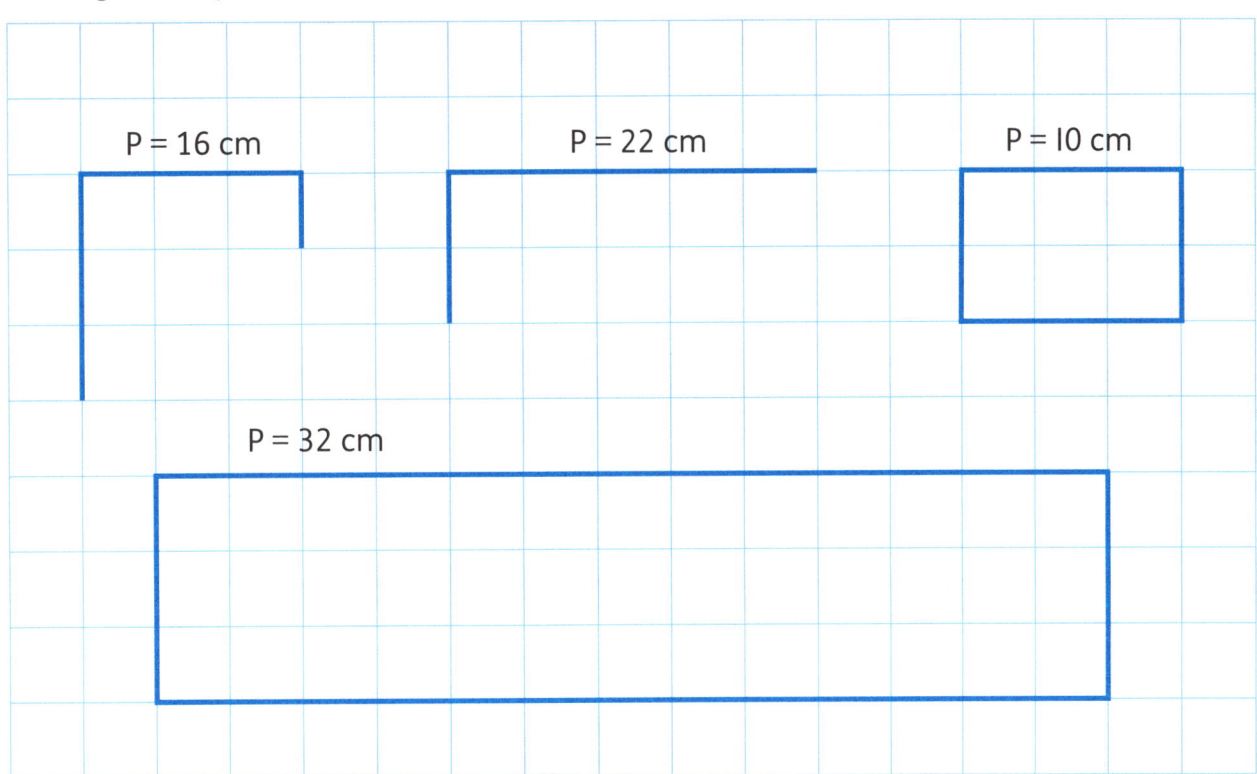

P = 16 cm   P = 22 cm   P = 10 cm

P = 32 cm

2 Draw any 10-sided shape with a perimeter of 34 cm.

▶ *Pupil Book page 100*

Area

> Area is the amount of space inside or covered by a 2D shape.

1 These shapes are drawn on a 1-cm grid. Which of these shapes has an area of 10 cm²? _____

Draw two different shapes on the grid, each with an area of 10 cm².

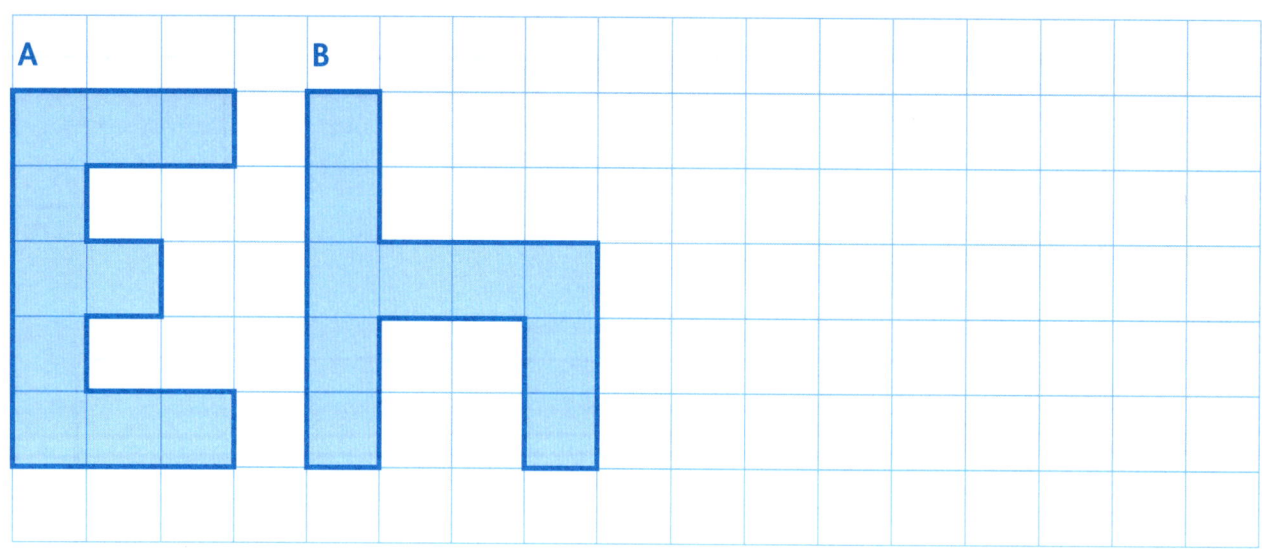

2 Draw three different rectangles with an area of 12 cm².

➡ *Pupil Book page 101*

More area

1 Work out the area of each shape and write it on the shape. One square on the grid represents 1 cm².

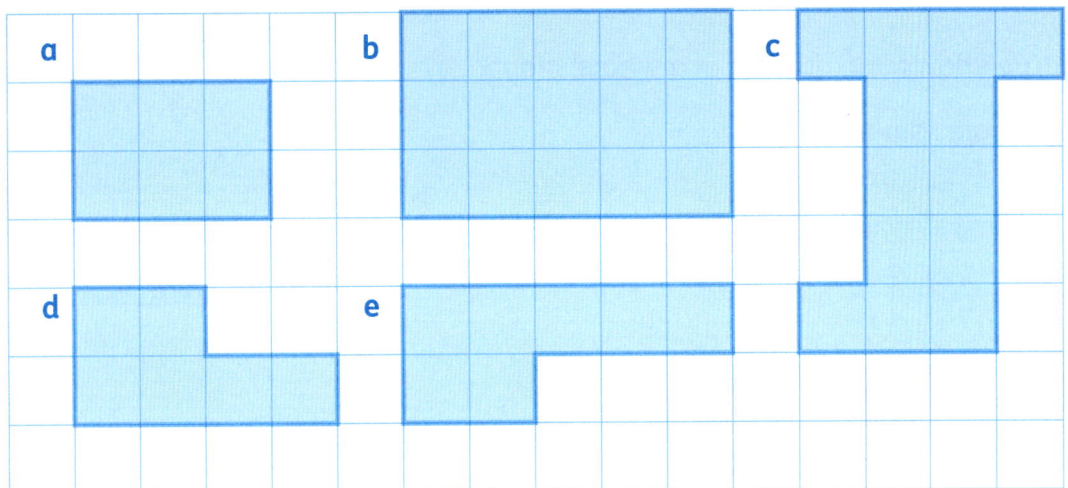

2 Some of the 1-cm squares on these shapes have been erased.

a Work out the area of each shape.

b Show your partner how you worked out how many squares there would be in each shape.

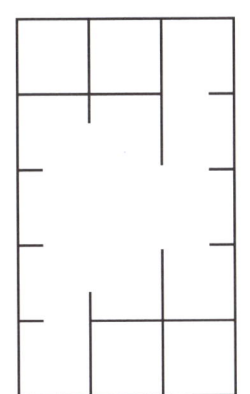
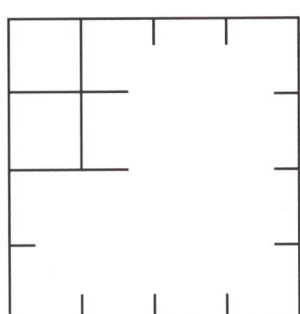
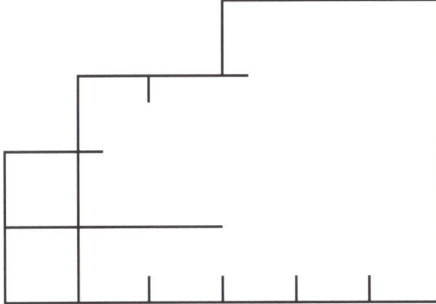

A   Area = _____

B   Area = _____

C   Area = _____

 Problem solving

Let the grid represent squares with 1-m sides.

3 Nisha uses 18 metres of rope to mark out a rectangle on the ground. The side lengths are all whole numbers and the rectangle covers the greatest possible area. Draw a sketch to show the lengths of the sides of the rectangle she marked out.

➡ *Pupil Book page 102*

Fractions

Revisit fractions (1)

1 Each picture shows one quarter of a whole shape.

Draw the whole shape.

a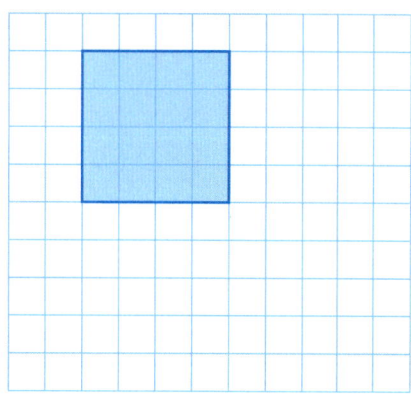

b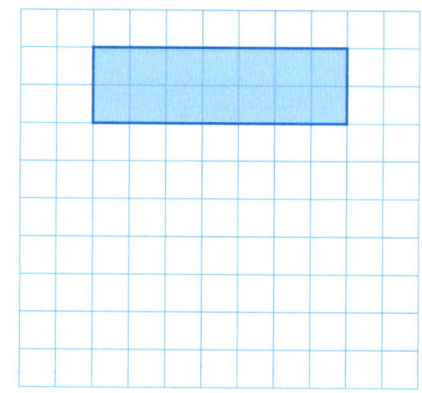

c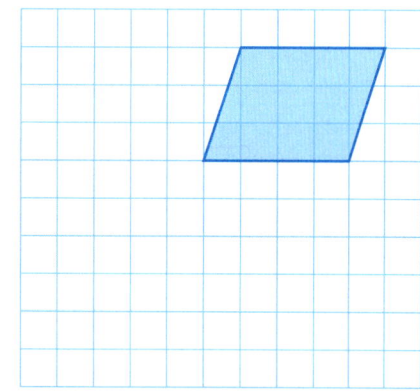

d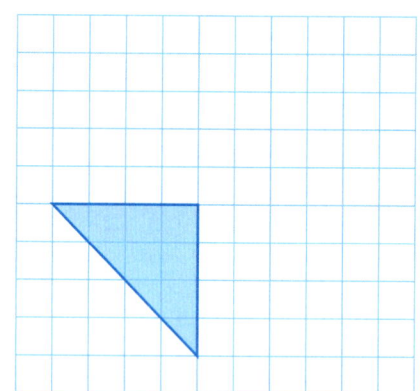

e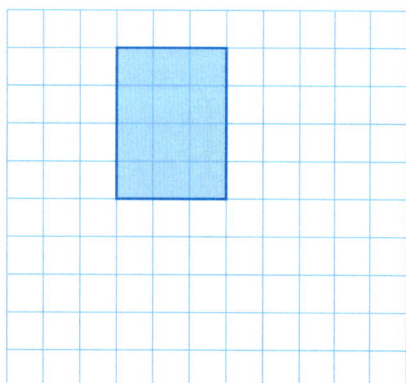

f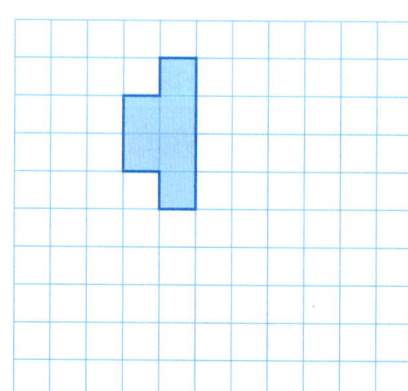

➡ Pupil Book page 103

Revisit fractions (2)

1 What fraction is shaded?

Circle the correct fraction.

a

| $\frac{1}{4}$ | $\frac{1}{5}$ | $\frac{4}{5}$ | $\frac{1}{6}$ |

b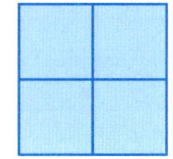

| $\frac{3}{4}$ | $1\frac{1}{4}$ | $1\frac{3}{4}$ | $2\frac{3}{4}$ |

c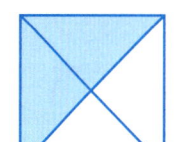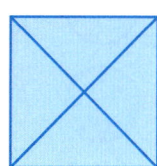

| $\frac{2}{3}$ | $\frac{1}{4}$ | $1\frac{2}{4}$ | $\frac{2}{4}$ |

2 What fraction of each set is shaded? Circle the correct fraction.

a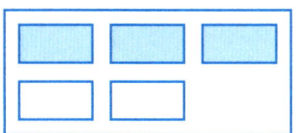

| $\frac{2}{5}$ | $1\frac{2}{5}$ | $\frac{3}{5}$ | $\frac{1}{5}$ |

b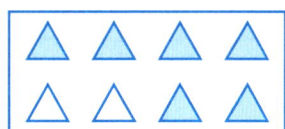

| $\frac{4}{8}$ | $\frac{6}{8}$ | $\frac{3}{8}$ | $\frac{2}{8}$ |

c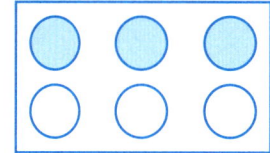

| $\frac{3}{6}$ | $\frac{1}{6}$ | $\frac{3}{4}$ | $\frac{1}{5}$ |

3 What is the missing numerator? Choose the correct number from the box.

 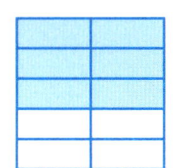

$$\frac{3}{5} = \frac{\square}{10}$$

| 6 | 8 |
| 4 | 3 |

4 What is the missing denominator? Choose the correct number from the box.

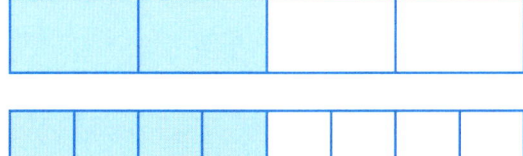

$$\frac{2}{4} = \frac{4}{\square}$$

| 5 | 4 |
| 6 | 8 |

5 Colour the given fraction of each set of shapes.

a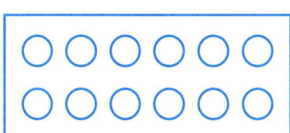

$\frac{7}{12}$

b

$\frac{1}{3}$

c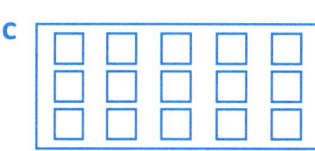

$\frac{2}{5}$

➡ *Pupil Book page 103*

Equivalent fractions (1)

Equivalent fractions have the same value.

These pictures show equivalent fractions.

 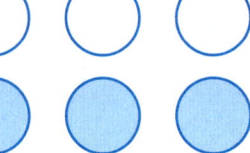

$\frac{1}{2} = \frac{3}{6}$

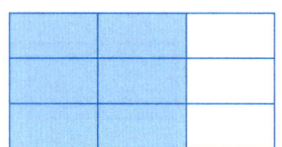

$\frac{2}{3} = \frac{6}{9}$

---

1 Write the equivalent fractions shown in each pair of pictures.

a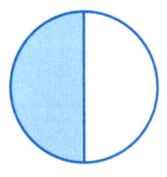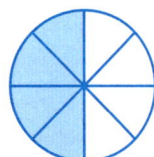

_____ = _____

b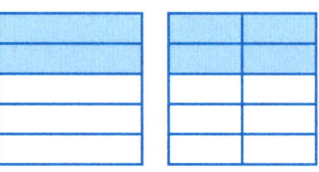

_____ = _____

c

_____ = _____

d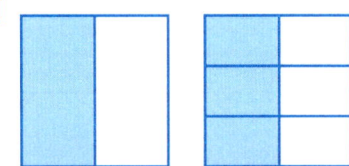

_____ = _____

e

_____ = _____

f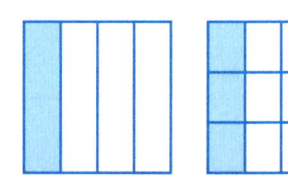

_____ = _____

2 Write the fraction shown in each picture and give an equivalent fraction.

a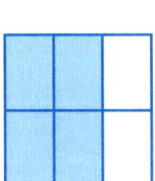

_____ = _____

b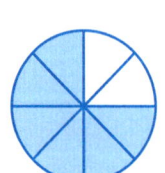

_____ = _____

c

_____ = _____

3 Draw your own fraction shapes, and write equivalent fractions for them.
You can divide up these shapes into equal sections to make your fractions.

a

_____ = _____

b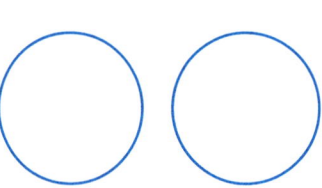

_____ = _____

c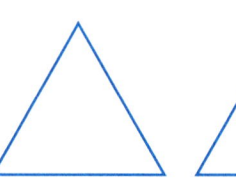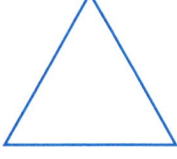

_____ = _____

➤ *Pupil Book page 105*

Equivalent fractions (2)

1 Fill in the missing fractions or numbers on each number line.

a

0          $\frac{1}{2}$ ☐

b

0   ☐   ☐ $\frac{3}{4}$ 1

c

0   ☐ ☐ $\frac{3}{8}$ ☐ ☐ ☐ $\frac{7}{8}$ 1

2 Use number lines to help you to complete the equivalent fractions.

a $\frac{1}{4} = \frac{\square}{8}$

b $\frac{1}{2} = \frac{\square}{4}$

c $\frac{6}{8} = \frac{\square}{4}$

d $\frac{8}{8} = \frac{\square}{4} = \frac{\square}{\square}$

3 Colour each shape to match the fraction given.

For each one write an equivalent fraction.

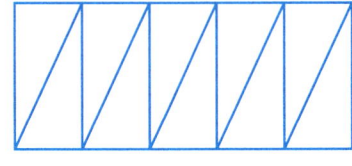      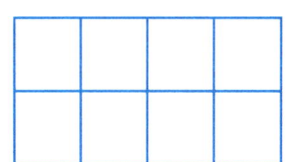

$\frac{4}{5} = \underline{\hspace{2cm}}$ $\frac{6}{8} = \underline{\hspace{2cm}}$ $\frac{1}{4} = \underline{\hspace{2cm}}$

4 Draw diagrams to show each pair of equivalent fractions.

a $\frac{1}{2} = \frac{2}{4}$ **b** $\frac{3}{5} = \frac{6}{10}$ **c** $\frac{2}{3} = \frac{8}{12}$

➡ Pupil Book page 105

Fractions and equivalent decimals

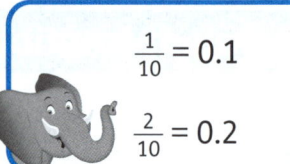

$$\frac{1}{10} = 0.1$$

$$\frac{2}{10} = 0.2$$

1 How much is shaded? Write your answer as a decimal.

a

b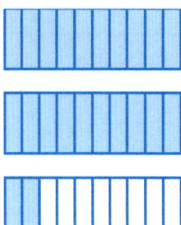

c

_____   _____   _____

2 How much is shaded? Write a fraction and a decimal.

a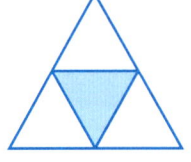

b

c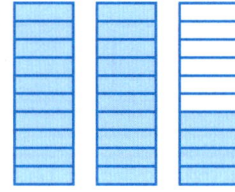

_____   _____   _____

3 Colour the given fraction of each 10 × 10 square.

a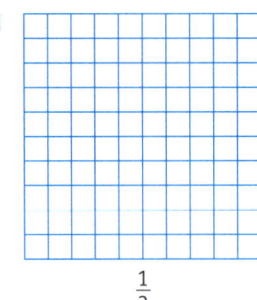

$\frac{1}{2}$

b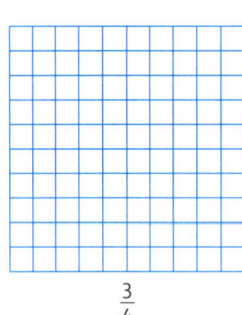

$\frac{3}{4}$

c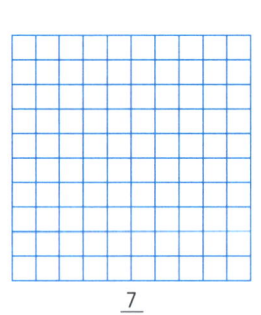

$\frac{7}{10}$

d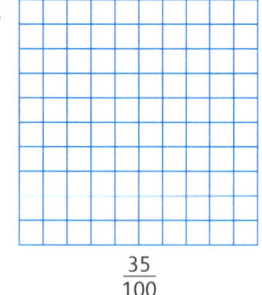

$\frac{35}{100}$

4 Mark the position of each fraction you shaded in question 3 on this number line.

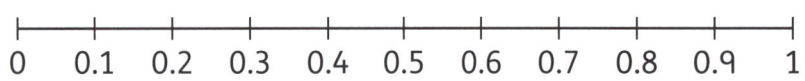

5 Draw lines to match each decimal to the equivalent fraction.

0.73   0.8   0.35   0.5   0.99   0.03   0.2

$\frac{35}{100}$ $\frac{20}{100}$ $\frac{8}{10}$ $\frac{73}{100}$ $\frac{4}{5}$ $\frac{3}{100}$ $\frac{1}{2}$ $\frac{99}{100}$

➡ *Pupil Book page 106*

Mixed numbers

A mixed number has a whole number part and a fraction part.
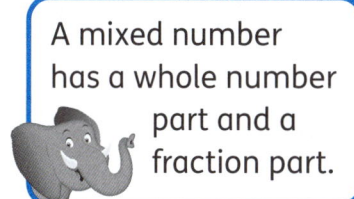

1 For each set, write the fraction that is shaded as a mixed number.

a    ____

b   ____

c    ____

d  ____

e  ____

f  ____

2 Write each mixed number from question 1 in the correct position on this number line.

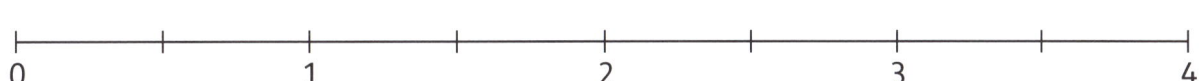

3 The mixed numbers in the box have fallen off the number line. Write the correct mixed number next to each arrow.

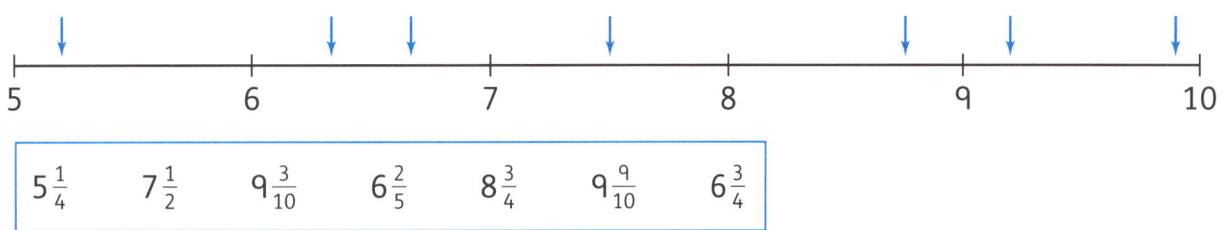

$5\frac{1}{4}$ $7\frac{1}{2}$ $9\frac{3}{10}$ $6\frac{2}{5}$ $8\frac{3}{4}$ $9\frac{9}{10}$ $6\frac{3}{4}$

4 Explain how you decided where each mixed number should go on the number line.

_____

_____

_____

➡ *Pupil Book page 107*

Write improper fractions as mixed numbers

In an improper fraction, the numerator (top number) is bigger than the denominator (bottom number).

$\frac{8}{3}$ is an improper fraction.

1 Colour each set of shapes to show the mixed number.

2 Write each mixed number as an equivalent improper fraction.

a $3\frac{1}{4} = \dfrac{\square}{\square}$

b $4\frac{3}{4} = \dfrac{\square}{\square}$

c $3\frac{5}{8} = \dfrac{\square}{\square}$

d $4\frac{1}{10} = \dfrac{\square}{\square}$

e $3\frac{7}{8} = \dfrac{\square}{\square}$

f $2\frac{4}{5} = \dfrac{\square}{\square}$

g $2\frac{8}{10} = \dfrac{\square}{\square}$

h $3\frac{2}{3} = \dfrac{\square}{\square}$

3 Fill in the missing numbers.

a There are _____ thirds in $2\frac{1}{3}$.

b There are _____ eighths in $3\frac{5}{8}$.

c There are 13 quarters in _____.

d There are 19 fifths in _____.

➡ *Pupil Book page 109*

Add and subtract fractions

1 Complete the target diagrams. The two numbers in each section must add up to the number in the centre. One section has been done for you.

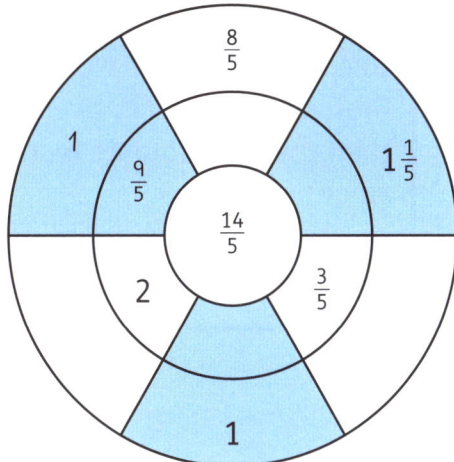 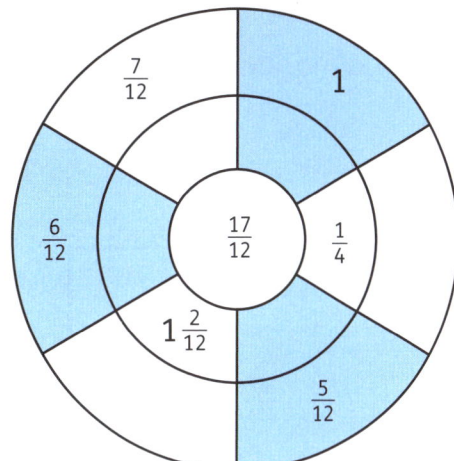

2 Colour each diagram to show the problem. Write the calculation to solve it.

a There are 20 cubes. Naresh takes some of these cubes to make a model.
He makes his model and has 8 of the cubes left over.

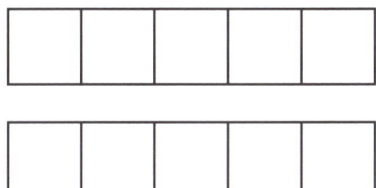

What fraction of the original cubes did he use for his model? _____

b Mia cuts two cakes into eighths so she has 16 pieces.
She eats 3 pieces and puts the rest out for her friends.
When her friends leave there is $\frac{1}{2}$ of one cake left.

What fraction of the two cakes did they eat?

_____

c Zara has two packs of 8 stickers. She uses $1\frac{1}{4}$ packs of stickers to decorate her book. Then she gives 3 stickers to a friend.

How many stickers are left?

_____

➡ *Pupil Book page 110*

Position and movement

Use coordinates

The grid shows the layout of a marine park.

 In coordinates, the *x*-coordinate (horizontal value) is given first, then the *y*-coordinate (vertical value).

Point a has coordinates (1, 2).

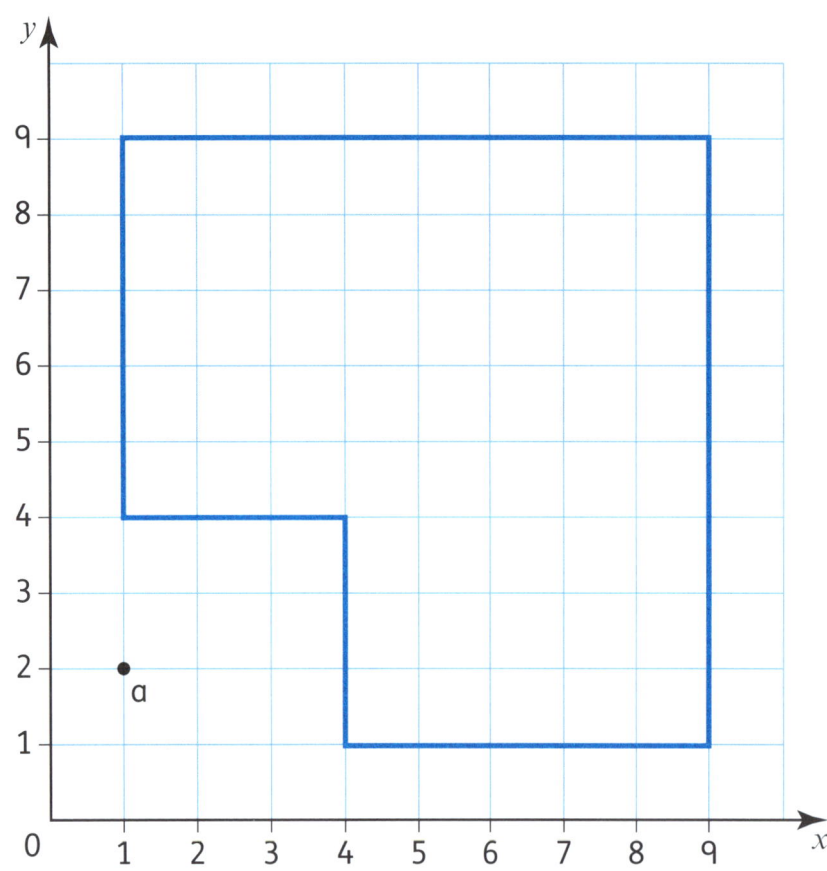

1. The coordinates of different places in the marine park are given.

   Add each place to the grid. Draw a dot and the letter to show each one.

   The first one has been done for you.

   a parking (1, 2)
   b entrance A (4, 1)
   c entrance B (2, 4)

   d café (8, 8)
   e bathrooms (6, 2)
   f bank machine (8, 2)

   g gift shop (5, 4)
   h shark tank (6, 3)
   i coral garden (4, 6)

   j turtles (2, 8)
   k anemone reef (2, 6)
   l tropical fish (4, 8)

   m eel cave (6, 8)
   n pelicans (8, 4)
   o starfish (8, 6)

2. Where would you put the octopuses? Add them to the grid and write the coordinates here:

   octopuses _____

▶ *Pupil Book page 115*

Translations

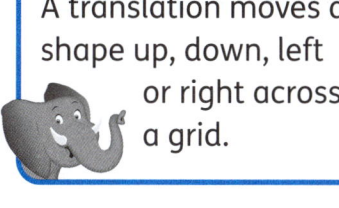

1 Follow the instructions to translate Shape A.

Redraw the shape in its new position.

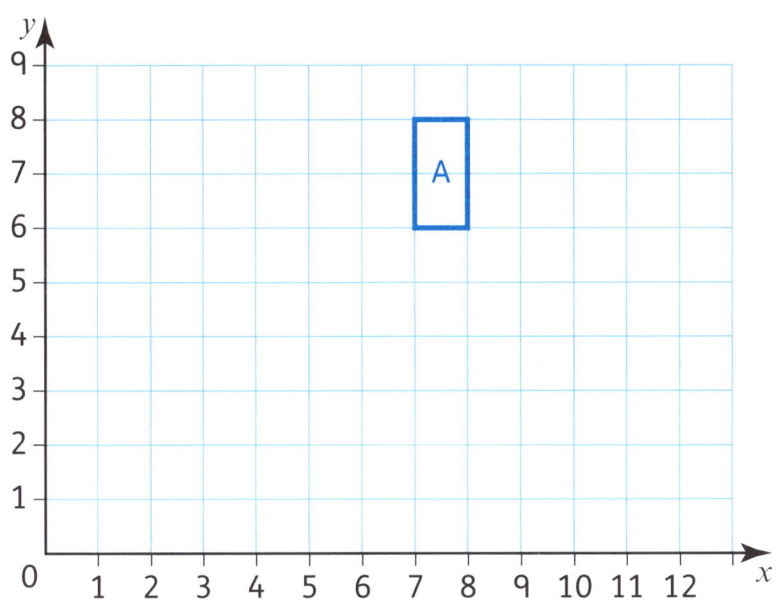

Start with shape A each time.

a 2 squares down, 2 squares right, label B   **b** 3 squares down, label C

c 7 squares left, 1 square up, label D   **d** 1 square up, 1 square right, label E

e 3 squares right, 1 square up, label F   **f** 4 squares down, 5 squares right, label G

2 Describe the translation that will move Shape G back to Shape A.

_____

Problem solving

3 Mandla translated Point A a total of 8 squares and Point B a total of 15 squares.

Work out a possible starting position for each point.

Show it on the grid.

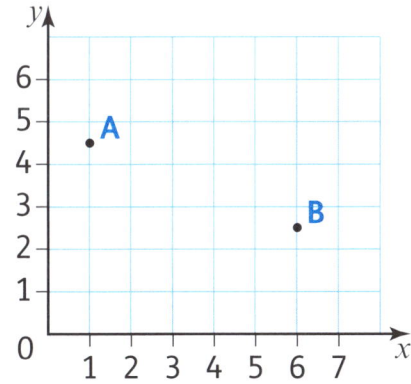

Use a pencil and make rough sketches to help you.

➡ *Pupil Book page 116*

Shapes on a grid

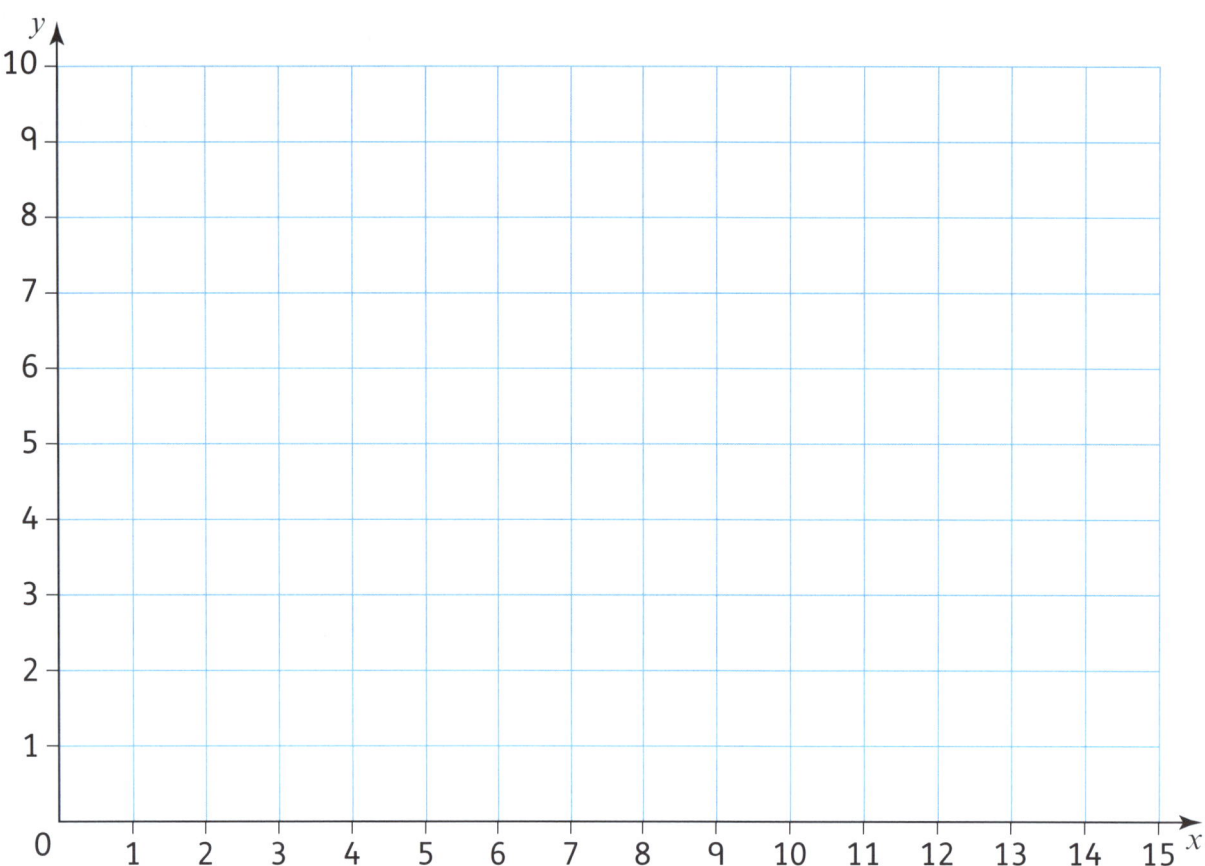

1 Plot and label each set of points.

Draw lines to join the points up in order, ending by joining the last point to the first point.

Write the name of each shape.

a   A(1, 10), B(4, 7), C(1, 7) _____

b   D(12, 10), E(14, 10), F(14, 3), G(12, 3) _____

c   H(8, 10), I(11, 7), J(8, 4), K(5, 7) _____

d   L(2, 1), M(6, 1), N(6, 3), O(4, 5), P(2, 3) _____

e   Q(9, 3), R(9, 0), S(13, 0) _____

f   Draw a five-sided shape of your own on the grid.

Label the vertices V, W, X, Y and Z.

Write the coordinates of each vertex.

V _____   W _____   X _____   Y _____   Z _____

▶ *Pupil Book page 117*

1   a   Join four of the points on the grid to form a quadrilateral.

     b   The dashed vertical line on the grid is a mirror line. Draw the mirror image of the quadrilateral on the other side of the line.

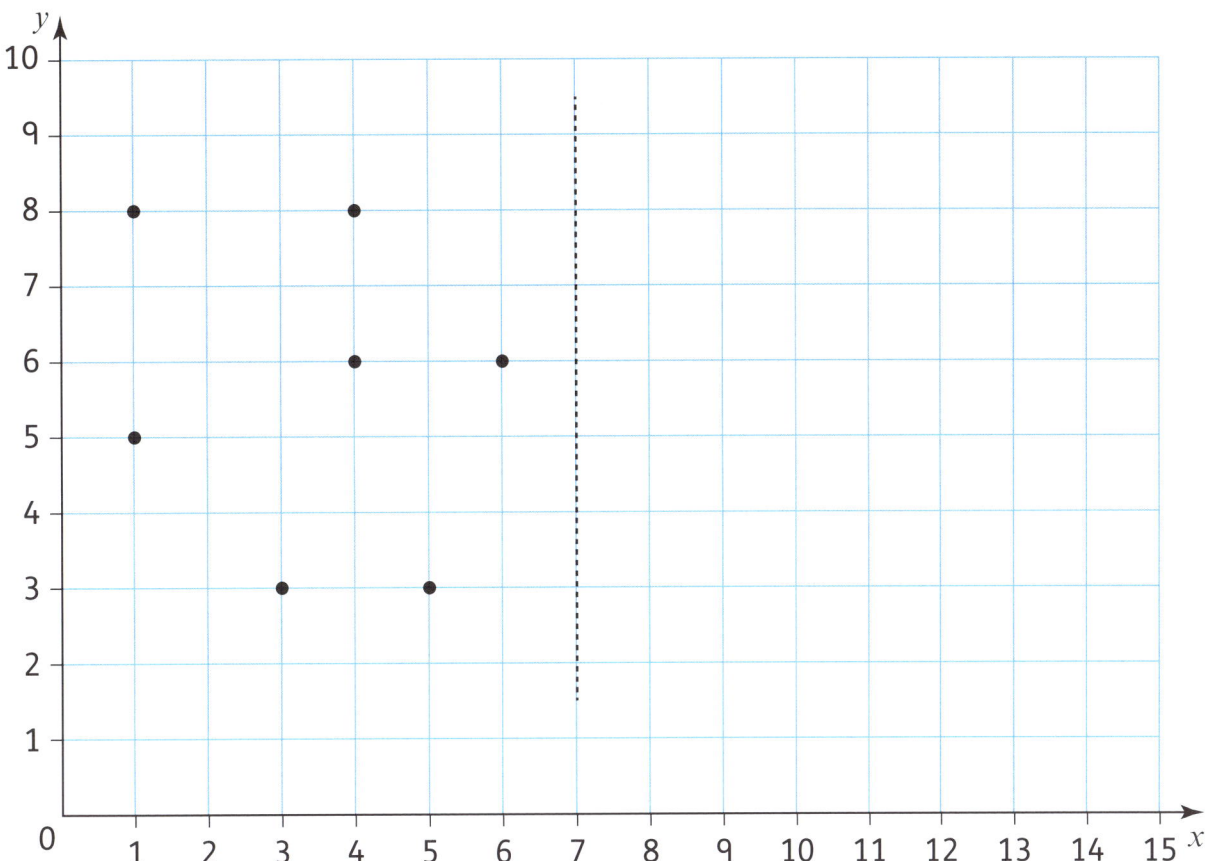

💡 **Problem solving**

Work in pencil and check your ideas using a mirror.

2   Half of a symmetrical shape has been drawn on the grid. Use what you know about symmetry to plot the vertices of the other half of the shape. Join up the lines to complete the shape.

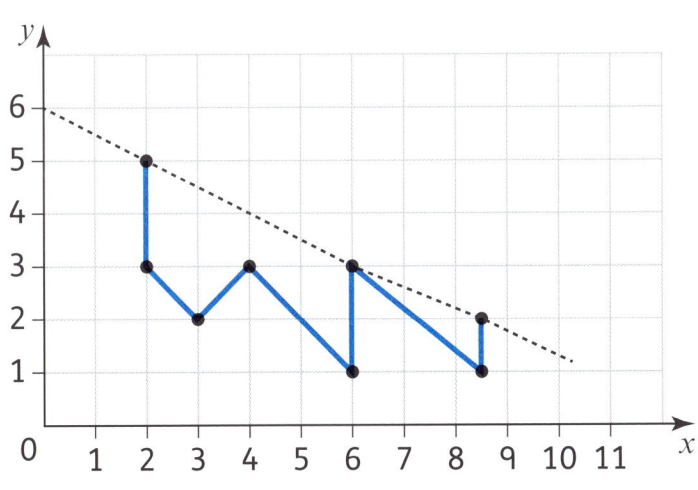

➡ *Pupil Book page 117*

Multiplication

Remember your facts

Complete each set of cards.

5 × 7 7 × 5	5 lots of seven	35

			54

	3 × 12 12 × 3		

	9 × 7 7 × 9		

			48

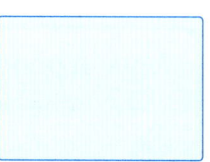

		3 lots of 8	

	6 × 4 4 × 6		

➡ *Pupil Book page 118*

Multiply larger numbers by 10

1 Fill in the missing numbers to complete each chain.

a (5) × 10 → ( ) × 10 → ( )

Remember:
8 × 10 = 8 tens = 80
18 × 10 = 18 tens = 180
123 × 10 = 123 tens = 1230

b (23) × 10 → ( ) × 10 → ( )

c (77) × 10 → ( ) × 10 → ( )

d (39) × 10 → ( ) + 5 → ( ) × 10 → ( )

e (121) × 10 → ( ) − 1000 → ( ) × 10 → ( )

f (426) × 10 → ( ) − 4000 → ( ) × 10 → ( )

g (509) × 10 → ( ) − 5000 → ( ) × 10 → ( )

2 This recipe makes enough cake for 6 people. Daniel is having a party for 60 people. How much of each ingredient does he need to make enough cake for all 60 people?

Vanilla Cake for 6 people	**Vanilla Cake** for 60 people
500 g sugar	_____ g sugar
4 eggs	_____ eggs
650 g flour	_____ g flour
250 ml milk	_____ ml milk
175 ml vegetable oil	_____ ml vegetable oil
12 g baking powder	_____ g baking powder
5 ml vanilla	_____ ml vanilla

➡ *Pupil Book page 120*

Multiply by 100

There are 100 centimetres in 1 metre.
To change a measurement from metres to centimetres, you multiply it by 100.

1 Write each of these metre lengths in centimetres.

a 4 m = ☐ cm

b 34 m = ☐ cm

c 73 m = ☐ cm

d 99 m = ☐ cm

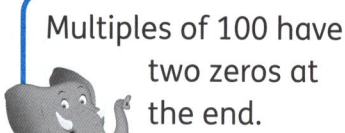

Multiples of 100 have two zeros at the end.

2 How tall is each tree in centimetres?

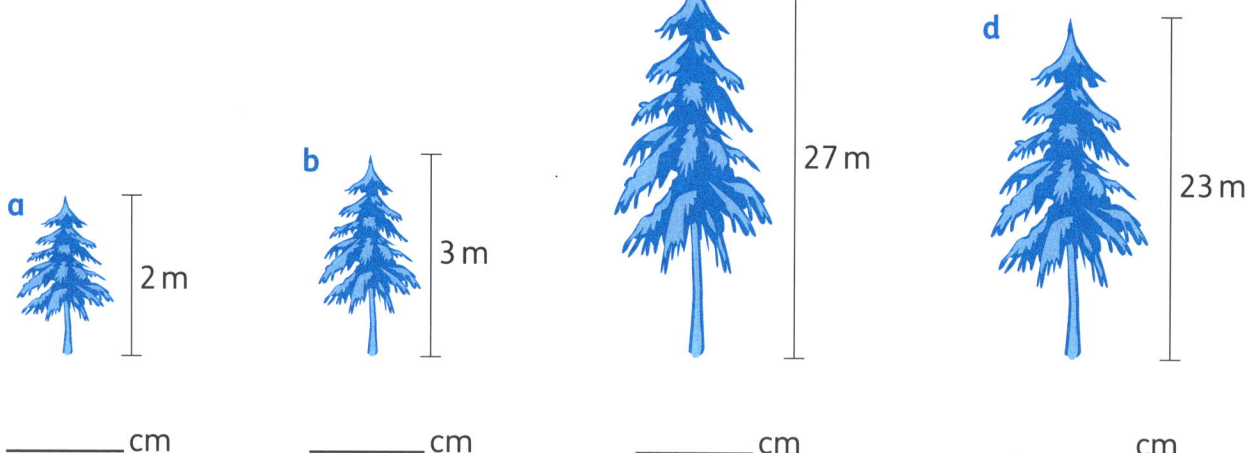

a 2 m

b 3 m

c 27 m

d 23 m

_____ cm _____ cm _____ cm _____ cm

💡 Problem solving

3 Sam stitched a tapestry 1 m long and 0.6 m wide.

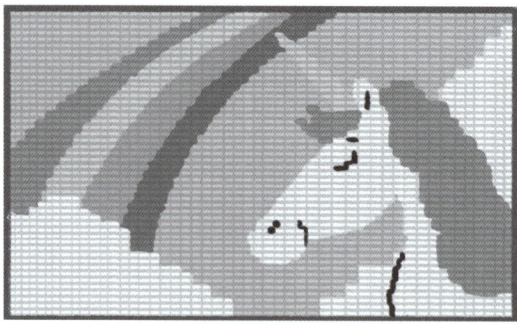

Draw a diagram or use patterns to help you.

In each row and column of stitches there are 10 stitches per centimetre.
How many stitches are around the perimeter of the tapestry? _____

➡ *Pupil Book page 121*

Multiplication challenge

Rows go across. Columns go from top to bottom.

Colour each row the correct colour.

25	55	75	15	35	65	red
29	59	79	19	39	69	yellow
22	52	72	12	32	62	blue
20	50	70	10	30	60	green
26	56	76	16	36	66	purple

1 a Multiply the numbers in the blue row by 3. ____ ____ ____ ____ ____ ____

   b Multiply the numbers in the yellow row by 6. ____ ____ ____ ____ ____ ____

2 a Multiply the green row by 10. ____ ____ ____ ____ ____ ____

   b What happens to the numbers? _____

   c Double the red numbers. What do you notice? ____ ____ ____ ____ ____ ____

   _____

3 a Multiply the purple numbers by 4. ____ ____ ____ ____ ____ ____

   b What do you notice about the last digit of each number? _____

   _____

4 Fill in the missing calculation in each diagram. Then work out the product.

   The first one has been done for you.

   a

   8×63

   8×60 8×3

   480   +   24

   Product: 504

   b

   7×40 7×8

   ____   ____

   Product: ____

   c

   9×86

   9×6

   ____   ____

   Product: ____

➡ *Pupil Book page 122*

More multiplying

Remember what you learnt about multiplying by 0.
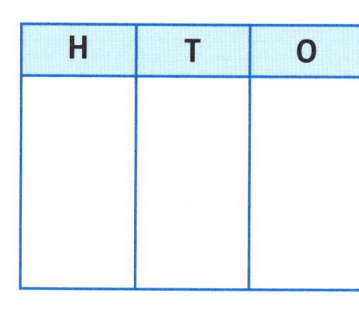

1 Anita used place-value charts to help her multiply 3-digit numbers.

Use the same method to complete each calculation.
The first one has been done for you.

a

H	T	O
••••	••	•••• / •••
••••	••	• ••• / •••
••••	••	•••• / •••

1200   + 60   + 21

```
   427
×    3
 1281
```

b

H	T	O

```
   343
×    6
```

c

H	T	O

```
   709
×    4
```

2 Complete each calculation using the digits on the cards.
The first one has been done for you.

a

```
[8] [0] [5]
×       [3]
  2  4  1  5
```

b

```
[ ] [ ] [ ]
×       [ ]
  4  0  1  5
```

c

```
[ ] [ ] [ ]
×       [ ]
  2  5  5  0
```

3 Write your own word problem to match each calculation in question 2.

a _____

_____

b _____

_____

c _____

_____

➡ *Pupil Book page 124*

Work with line graphs

Draw and interpret line graphs

1 This table gives the temperature in a classroom during a school day.

Time	8.00	9.00	10.00	11.00	12.00	13.00	14.00	15.00
Temperature	22	24	26	27	27	28	28	26

Draw a line graph to show this data. Make sure you include a title.

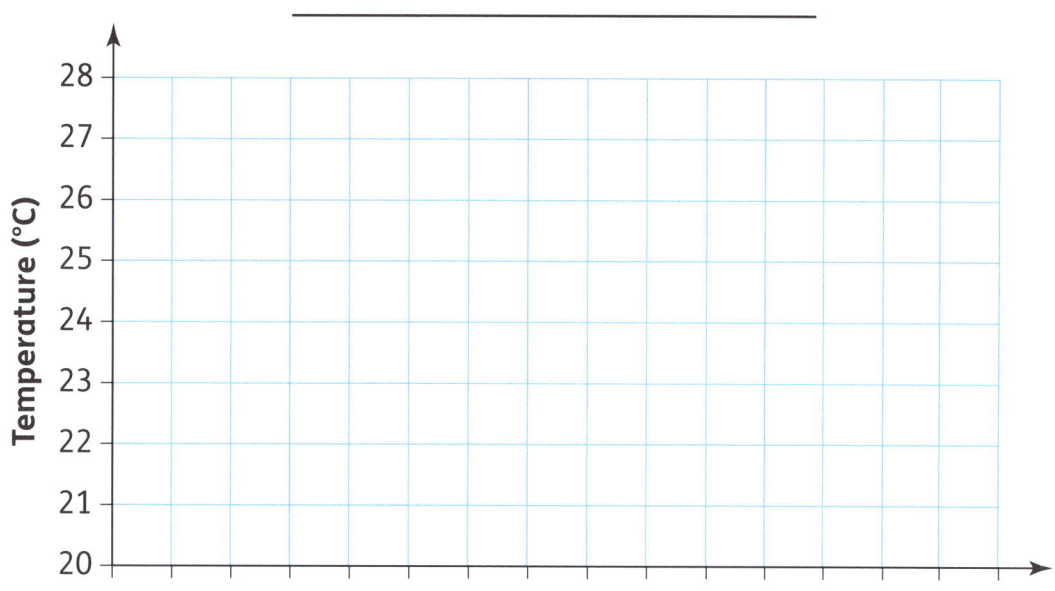

Time

2 Use your graph from question 1 to answer these questions.

a What temperature was it at 10 a.m.? _____

b What happened to the temperature between 1 p.m. and 2 p.m.?

_____

c Why does the line go down from 14.00 to 15.00?

_____

d Estimate when you think the temperature in the classroom reached 25 °C.
Explain how you worked this out from the graph.

_____

_____

➡ *Pupil Book page 129*

Read and interpret line graphs

1 The graph shows the length (height) of a baby measured in centimetres each month from birth to 6 months.

a Use the information on the graph to complete the table.

Age in months	Length in centimetres
1	
2	
3	
4	
5	
6	

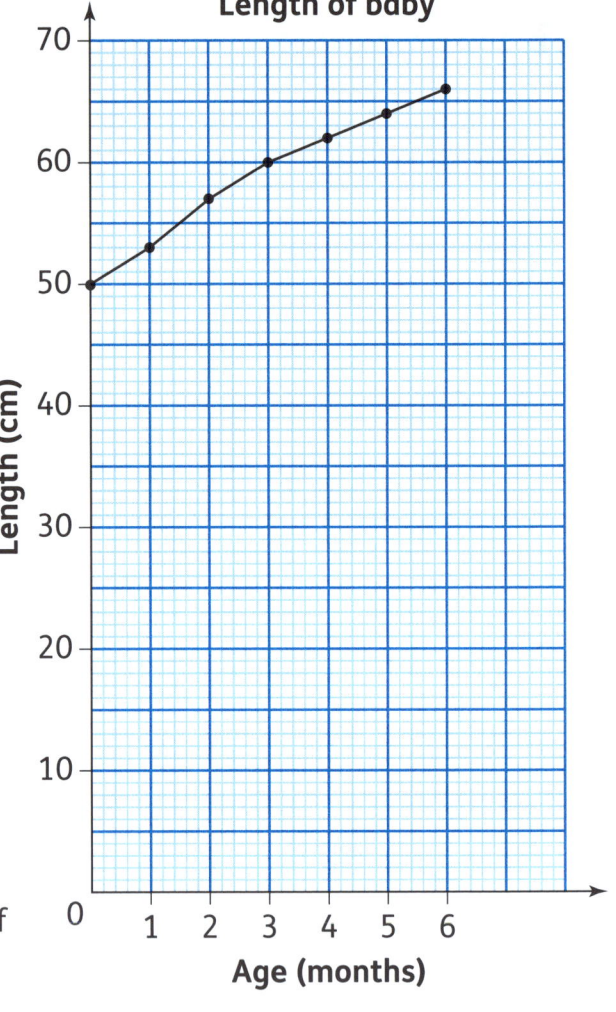

Length of baby

b At the age of 7 months, the baby's length is 67 cm. Add this data to the graph.

c How many centimetres did the baby grow between month 1 and month 2?

_____

d What is the difference between the length of the baby in month 3 and month 7?

_____

e When did the baby's length increase the most?

Between month _____ and month _____

f What does the graph tell you about the baby's growth between month 4 and month 6?

_____

g At the end of month 8, the baby has grown another 1.5 cm.

Show this on the graph.

What is the length of the baby in metres at the end of month 8?

_____

➡ *Pupil Book page 130*

More line graphs

Anjali swims every day to train for a gala. At the end of each week, she records how long it took her to swim 50 metres (to the nearest second).

Some of her data is on the graph. Some of it is in the table.

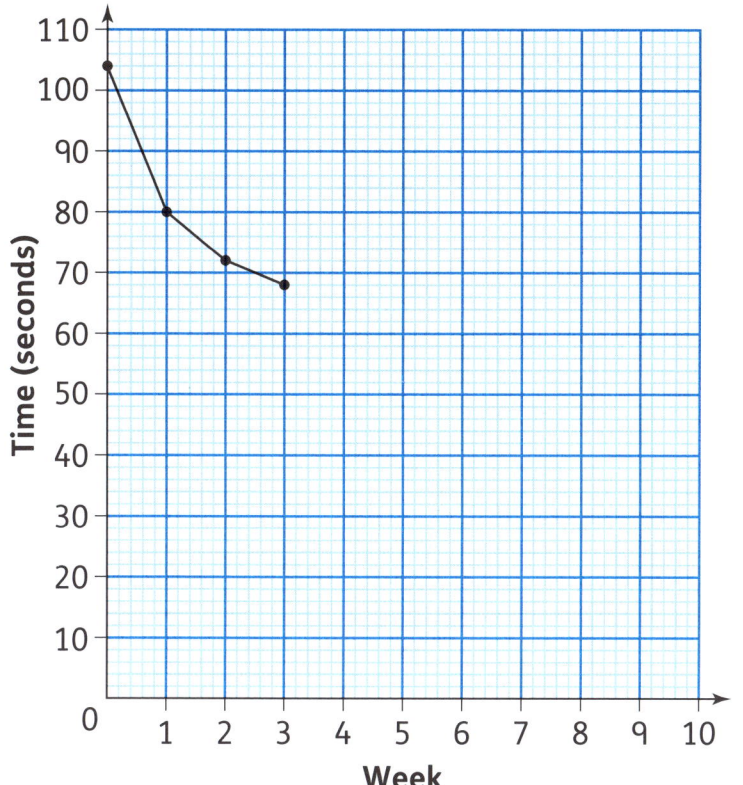

Week	Time (s)
1	
2	
3	
4	62
5	68
6	64
7	58
8	52
9	52
10	52

1 **a** Use the data on the graph to fill in the missing rows in the table.

b Use the data in the table to complete the graph.

2 Answer these questions about Anjali's data.

a What does the line on the graph tell you about her progress over ten weeks?

_____

_____

b How long did it take Anjali to halve her original time?

_____ weeks

c How much faster was Anjali's swim time for 50 m at the end of week 10 than at the end of week 1?

It was _____ seconds faster.

d Do you think she will get much faster if she keeps training? Give a reason for your answer.

_____

_____

➡ *Pupil Book page 130*

Patterns

Investigate patterns and rules

> A sequence is a number pattern that follows a rule to get from one term to the next.
>
> 3, 5, 7, 9
>
> The term-to-term rule for the sequence is 'add 2'.

1 Mthunzi made these number patterns by counting on or back in steps of the same size.

Work out the term-to-term rule he used for each pattern. Fill in the missing numbers.

a 142,   146,   150,   ____, ____, ____

b 120,   125,   130,   ____, ____, ____

c 80,   120,   160,   ____, ____, ____

d ____, ____, 81,   84,   87,   ____, ____

e ____, ____, 148,   145,   142,   ____, ____

2 Write out the term-to-term rule for each of these number sequences. Use the rule to find the next three terms. The first one has been started for you.

a 125,   130,   135,   140,   ____, ____, ____

   Rule: _____Add 5_____

b 520,   620,   720,   ____, ____, ____

   Rule: _____

c 999,   899,   799,   699,   ____, ____, ____

   Rule: _____

d 179,   169,   159,   149,   ____, ____, ____

   Rule: _____

3 Follow the term-to-term rule for each sequence to write the next four terms.

a Add 100   423, ____, ____, ____, ____

b Subtract 10   325, ____, ____, ____, ____

c Subtract 5   185, ____, ____, ____, ____

d Add 50   230, ____, ____, ____, ____

e Subtract 4   12,   ____, ____, ____, ____

➡ *Pupil Book page 131*

Square numbers

This square shows 2 squared. Each side is 2 squares long. 2 squared covers 4 small squares.

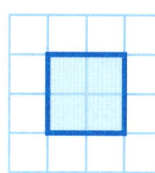

This is 4 squared. Each side is 4 squares long. 4 squared covers 16 small squares.

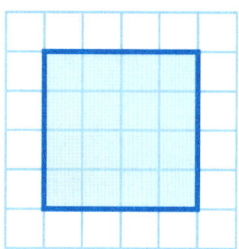

1 Shade squares on the grid to square the numbers in a – c.

a 5 squared = _____ **b** 3 squared = _____ **c** _____ squared = 64

2 Complete each number fact.

a _____ squared = 4 **b** 7 squared = _____ **c** _____ squared = 81

➡ *Pupil Book page 133*

Division

More division

There are different ways to record your work when you divide.

43 ÷ 4	51 ÷ 3
43	17
−40 10 × 4	3)‾51
───	30 10 × 3
3	──
	21
43 ÷ 4 = 10 remainder 3	21 7 × 3
	──
	0 17 groups of 3
	No remainder

1 8 bottles fit into a box.

How many boxes can I fill with:

a 48 bottles _____

b 55 bottles _____

c 60 bottles _____

d 84 bottles _____

e 100 bottles? _____

Write down the number of full boxes and any bottles left over.

2 5 people fit into a car.

How many cars do we need for:

a 34 people _____

b 45 people _____

c 69 people _____

d 85 people _____

e 93 people? _____

Every person has to be in a car. One of the cars may not be full.

3 Use facts you know to find the answers to these divisions.

a 15 ÷ 3 = _____   b 18 ÷ 3 = _____   c 18 ÷ 6 = _____

d 16 ÷ 4 = _____   e 21 ÷ 3 = _____   f 32 ÷ 8 = _____

g 40 ÷ 8 = _____   h 45 ÷ 9 = _____   i 80 ÷ 8 = _____

j 90 ÷ 9 = _____   k 50 ÷ 10 = _____   l 15 ÷ 5 = _____

m 25 ÷ 5 = _____   n 20 ÷ 10 = _____   o 36 ÷ 6 = _____

➡ *Pupil Book page 135*

Divide by 10 and 100

1. Fill in the missing numbers in each function machine.

a

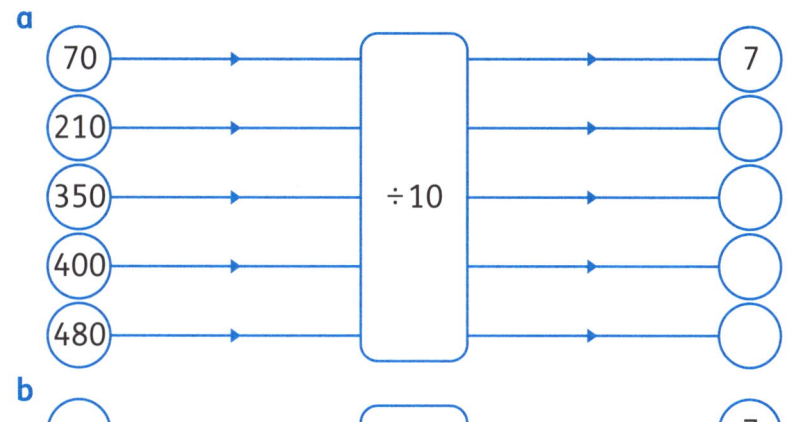

Input		Output
70	÷10	7
210		
350		
400		
480		

When you divide by 10, the digits move one place to the right.

$70 \div 10 = 7$

b

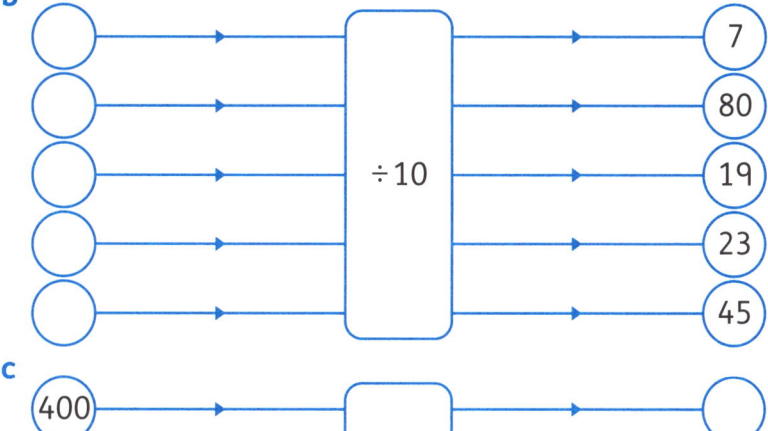

Input		Output
	÷10	7
		80
		19
		23
		45

c

Input		Output
400	÷100	
1600		
		18
		32
		9

When you divide by 100, the digits move two places to the right.

$300 \div 100 = 3$

Problem solving

2. Mike has a model car. It is 10 times smaller than a real car. If the model is 42 cm long, how long is the real car?

_____

_____

3. A picture of a ruler is $\frac{1}{10}$ of the size of a real ruler. The real ruler is 30 cm long. How long is the ruler in the picture?

_____

_____

➡ *Pupil Book page 136*

Divide or multiply?

1 Fill in the missing operation sign (× or ÷) to make each number sentence correct.

a 40 ☐ 5 = 8

b 6 ☐ 8 = 48

c 6 ☐ 6 = 1

d 6 ☐ 6 = 36

e 100 ☐ 10 = 10

f 10 ☐ 10 = 100

g 9 ☐ 9 = 81

h 81 ☐ 9 = 9

i 300 ☐ 10 = 30

j 5 ☐ 6 = 30

k 7 ☐ 8 = 56

l 200 ☐ 10 = 20

Problem solving

2 Can you solve these number riddles? Draw lines to match the riddles to the correct numbers in the box.

a When I am multiplied by 3 I make 18.

b If I am 5 times larger I make 45.

c When I am divided into 3 equal groups, there are 7 in each group.

d When I am doubled I make 36.

e If you divide me by 10 you get 3.

f If you divide me by 9 you get 9.

3
6
9
18
21
27
30
60
81

3 Make up your own riddles for the numbers that are left over.

_____

_____

_____

_____

_____

➡ *Pupil Book page 137*